Failure Mechanisms

— IN —
BUILDING CONSTRUCTION

EDITED BY DAVID H. NICASTRO

Published by

ASCE
PRESS

American Society of Civil Engineers
1801 Alexander Bell Drive
Reston, Virginia 20191-4400

Abstract:

This book is the result of the editor's efforts to understand why buildings fail. He has combined his experience investigating the causes of building failures with extensive research, in order to present detailed descriptions of the causes and identifying characteristics of a wide range of failure mechanisms. This book also includes an index of these various mechanisms, along with numerous case studies taken and expanded from his ongoing magazine column on "Failures," and an annotated bibliography. There is also a section on performing failure investigations to put the material into a practical context.

Library of Congress Cataloging-in-Publication Data

Failure mechanisms in building construction / David H. Nicastro, editor.
p.   cm.
ISBN 0-7844-0283-3
1. Building failures--Investigation.   I. Nicastro, David H. (David Harlan)
TH441.F35 1997   97-25452
690'.21--dc21   CIP

DEDICATION

To Susan, my endlessly patient wife, who encouraged me as this little task
continued to grow

CONTENTS

ILLUSTRATIONS

ACKNOWLEDGMENTS

This book was made possible by the many individual contributions of numerous engineers who have published in the scientific literature their case histories, research, and theoretical propositions on the nature of failure mechanisms. I appreciate their collective efforts to advance the art and science of engineering, and I hope I have accurately summarized their conclusions.

I received tremendous support from ASCE's Technical Council on Forensic Engineering, and its Committee on the Dissemination of Failure Information, on which I serve. Many of the ideas presented here were first discussed and studied in these groups.

The case histories presented in this book were first published as monthly columns under the title "Failures" in *The Construction Specifier*. I wish to thank my editors at the Construction Specifications Institute (CSI) for their assistance in both the original publication and this re-presentation.

Finally, I would like to thank my research assistant, Jeff Morris, for his thorough study of so many diverse topics in engineering that culminated in this publication.

INTRODUCTION

This book evolved from my curiosity about why things fail. I am endlessly fascinated with the discovery of the root causes of the physical behavior of building components. In my engineering practice investigating the causes of building failures, I have made many observations about the physical manifestations of distress and have hypothesized their causes. Because I had been exposed to such a wide variety of failure mechanisms in my career, I began to wonder if it would be possible to catalog all of them. No doubt such an endeavor would always be easier to do later in my career, but I was impatient to discover how many truly unique failure mechanisms there are.

This book presents the results of my research, organized into several sections:

1. Descriptions of failure mechanisms, with their causes and identifying characteristics, in a narrative format. I do not claim that this list is exhaustive, but my goal was to be as comprehensive as possible.
2. An index of these failure mechanisms in a tabular format, which can be used as a diagnostic tool.
3. Case histories, which are cross-referenced to the failure mechanisms in the tabular index.
4. A bibliography, citing references for most of the failure mechanisms discussed.

I have also included a section on performing failure investigations, to provide context for the appropriate use of this information.

A detailed understanding of failure mechanisms is essential to the practice of forensic engineering. I define "failure mechanism" as an identifiable phenomenon that describes the process or defect by which an item or system suffers a particular type of failure. Thus, "punching shear" is a failure mechanism because it is an identifiable phenomenon, but "shear" is not. (However, the broader term "failure mode" can encompass fundamental behavior, like shear.)

Although many cases hinge on the exact technical cause of failure, failure mechanisms are not always the most important aspect. In many cases, it is trivially obvious what caused the failure from a *technical* standpoint, but the focus of the case is the *procedural* cause of the failure—the circumstances that lead to the occurrence.

To use one of the most famous structural disasters as an illustration of this point, consider the collapse of the Hyatt Regency Hotel pedestrian walkways in Kansas City, Missouri in 1981. The investigators determined that the failure was initiated simply by overstressing a steel connection—the failure mechanism was nothing more mysterious than overloading a poor connection detail. What was more important in that case was the process by which the inappropriate connection detail was designed, subsequently modified, approved, and finally constructed.

Nevertheless, it is important to study failure mechanisms in forensic engineering and also in design engineering, the essence of which is to prevent failures; in fact, a useful working definition of engineering is to design against failure. To the extent that we can predict all possible failure mechanisms for a particular system, then we can design the system to resist each of those mechanisms. Therefore, a failure-resistant system is one in which all possible failure mechanisms are known; the design addresses each of these mechanisms; and the construction is executed consistent with the design.

Although the literature is replete with articles on individual failure mechanisms, what has been missing is a synthesis of this disparate information. This book attempts to fill that void.

Chapter 1. Performing Engineering Investigations

The essence of an engineering investigation of a construction failure is to determine the cause—not only the failure mechanisms but the procedural causes and contributing factors as well. While the majority of this book is devoted to the technical concepts of failure mechanisms, it would be a gross oversight not to consider the context in which most investigations are performed.

Forensic engineering is the application of engineering principles to the investigation of failures or other performance problems. The word "forensic" means public forum, which today usually means courtroom. In engineering, this term has various other meanings and connotations, especially related to failure analysis. The word forensic is best reserved for projects that involve dispute resolution services.

"Forensic" does not imply expert testimony; on the contrary, most forensic projects are resolved before providing testimony. It is convenient, however, that many people think that it implies testimony; whether litigation is ultimately involved or not, forensic services should be provided as if courtroom testimony will eventually be required.

Although negotiation, settlement, and alternative dispute resolution techniques might be used, the default method for resolution is litigation. Therefore, project activities must be undertaken with the understanding that testimony may ultimately be required just because there is a dispute involved; proper documentation is critical. Testimony will, in fact, only be required in a small percentage of the projects.

One of the primary criticisms of forensic engineering is that it is not productive for society. To counter this impression, the objective of reputable forensic engineering services is to facilitate fast, fair, and cost-effective resolutions of disputes. Currently, there is a trend toward using alternative dispute resolution (ADR) methods, which I encourage because these methods promote fast and cost-effective resolutions. Most commonly, ADR includes mediation and arbitration, but there are other methods recognized by the courts as well. Engineers can serve a role in ADR as well as litigation, by serving as a witness, arbitrator, or mediator.

It is important to recognize that cause does not automatically imply fault; culpability is a complex issue involving legal liability as well as causal actions. It is common in forensic projects to discuss culpability for the analyzed problems with clients and their legal counsel. An engineer, however, should resist providing testimony (either in court or deposition) related to culpability, which is generally not a professional engineering opinion. An exception is addressing whether another engineer has met the applicable standard of care in the performance of his/her work; this opinion requires a professional engineer.

Engineering analysis requires judgment, not judgmentalism. Some of the largest disasters were caused by the smallest, most innocent mistakes. It is easy to adopt a "Monday morning quarterback" attitude in forensic engineering, but humility is warranted; some failures are bound to occur in the career of any designer.

Unlike an attorney, an engineer should not be an advocate for his/her client in a construction dispute. It is inherent in the adversarial process that the engineer will become more familiar with the client's personnel and technical positions than with the opponent's; however, he/she must endeavor to remain impartial. It should not be discernible from the opinions one states which side of a controversy one is on. As a device for remaining impartial, before stating an opinion consider whether it would be stated differently if developed for the opposition. If so, then re-word the opinion so that it would not be different.

Because a conflict of interest is a professional ethics issue, it is important to avoid working (or even appearing to work) for two parties with opposing interests. Therefore, a conflict of interest review should be conducted before accepting a forensic project. This review should encompass current and previous clients and should include other parties involved in projects as well.

If a potential conflict of interest is discovered during these review procedures, emphasis should be placed on full disclosure and consent from all parties. If a conflict is discovered after a project is under way, the circumstances are more difficult, but full disclosure should be emphasized.

The following discussion is organized according to a typical scope of services for a forensic investigation.

Perform a Cursory Overview

Before getting too deeply involved in a case, it is prudent to make a cursory overview to determine whether you are qualified to be an expert in the case and whether enough facts exist to render substantiated opinions. This review should include a site visit and review of any available documents. This phase of work should culminate in a definition of the problem in precise engineering terms. It should also be possible after a cursory overview to make a preliminary estimate of your client's exposure to liability (for a defendant) or potential damages (for a claimant).

Perform Fieldwork

It is important to document the existing conditions at the time of failure with photographs, videotape, and field notes. While the fieldwork activities may be the same as for traditional projects, forensic projects require more careful documentation. Record all relevant parameters, not just gross observations. For example, to say "the concrete is cracked" is not nearly as useful as documenting the size, shape, spacing, pattern, and so forth.

Pay particular attention to temporal information. Many failure mechanisms can be ruled out simply on the basis of the time of occurrence or duration of a process. Generally, the development or progression of a failure mechanism over time will follow one of three general curves, as shown in Figure 1.1.

Time is plotted along the horizontal axis, and the number of occurrences or magnitude of distress is plotted along the vertical axis. Excluding those mechanisms that are instantaneous or sudden (such as brittle fracture), the remaining processes will generally fall into one of three relationships with time.

1. Roughly linear—mechanisms that occur at a generally constant rate, such as carbonation, and do not accelerate or decelerate with time.
2. Diminishing with time—mechanisms that are initial construction phenomena, such as

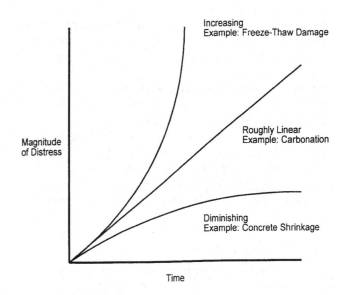

FIGURE 1.1 Time Dependence of Failure Mechanisms

concrete shrinkage, and occur at an ever-decreasing rate with time.
3. Increasing with time—mechanisms that are cumulative and accelerate with time as the previous distress exacerbates the process, such as freeze-thaw damage.

A common mistake of investigators is to attempt to develop hypotheses too early. They would be much better served to train themselves simply to observe first and develop theories later. In recording the observations, it is best to separate objective from subjective information. This will avoid the natural tendency to see and collect only the evidence that is consistent with a pet theory.

This process is consistent with the Scientific Method, which can be summarized as follows:

1. State the problem
2. Make observations
3. Form hypothesis
4. Perform test or experiment
5. Analyze outcome to develop conclusions
6. Iterate: form new hypothesis, test, and analyze

The tests of engineering hypotheses may include water infiltration testing, load tests, laboratory analysis, or trial remedies. It is important to remember that there are very few experiments that are conclusive for both outcomes; most tests yield "yes/maybe" or "no/maybe" outcomes, and therefore require iteration to achieve a definitive conclusion.

During the fieldwork, specimens can be collected for subsequent laboratory testing. They should be individually numbered and labeled and photographed before removal from their surroundings, to document their original conditions. If the specimens leave your possession, a chain of custody log should be used to track who had care, custody, and control of them at all times.

Perform Laboratory Testing

In most forensic projects, some laboratory testing is necessary; for example, microscopic measurements, determination of material properties, or chemical analysis. As much as possible, these tests should be performed in strict accordance with recognized industry standards, and any deviations should be reported.

Perform Research

For each component in a case, it is typically necessary to research industry standards, code compliance, design documents, possible failure mechanisms, or material behavior. For each subject attempt to answer the following questions, which break the issues into four distinct categories:

1. What did industry standards call for?
2. What did the design documents call for?
3. What was actually constructed?
4. What changed after construction?

In this phase, it is also often necessary to perform engineering calculations. Unfortunately, most available formulas were developed for design, not analysis. A common mistake by investigators is to algebraically manipulate an empirical design formula to solve for a variable other than the one intended by the original derivation of the equation. For example, the Uniform Building Code (UBC) provides the following formula for calculating wind load:

$$P = C_e C_q q_s I$$

where P = design wind pressure; C_e = combined height, exposure, and gust factor coefficient; C_q = pressure coefficient for the structure or portion of structure under consideration; q_s = wind stagnation pressure at the standard height of 30 ft; and I = importance factor. These coefficients and factors are provided in tables in the UBC.

While it is tempting to solve for any variable in an equation, clearly in this example it would be meaningless to solve for the importance factor I given a particular design wind pressure P and the other coefficients. The equal sign in an empirical formula is not the same as in algebra; the terms must remain on their original side of the equal sign.

Design formulas, however, can be useful starting points for listing the possible contributing factors to a failure. For example, consider a case involving excessive deflection of a structural member in which the designer's expected deflection is given by the following formula:

$$D = \frac{5}{384} \frac{WL^4}{EI}$$

where D = deflection; W = uniform load on the member; and L, E, and I = members' length, modulus of elasticity, and moment of inertia, respectively.

We can immediately list the potential errors that caused the deflection to exceed the designer's expectation, simply by examining the design equation.

- Is it the correct equation for the actual conditions? This particular formula would be appropriate for the midspan deflection of a uniformly loaded, simply supported beam.
- Is the load really uniform (W)? Most loads are not.
- Did the designer use the correct length (L), measuring from the actual points of support? Often there is a discrepancy between the assumed locations of fixity (such as the centerline of supporting members) and the actual locations.
- Are the section properties (I) and the material properties (E) the same as intended?

Formulate Opinions

Armed with all of the objective findings from the fieldwork, laboratory tests, and research, it is possible to formulate opinions that will stand up to the scrutiny of the adversarial process.

Correlation is the strongest tool available to a forensic engineer, so keen observation is necessary to discover links between findings. But beware of "hidden third causes." Two correlating observations may not have a causal relationship with each other; they may, in fact, both be the result of a third cause.

When rendering an opinion, it should be based on an accurate understanding of the degree to which it is supported by the evidence. In most projects, the findings do not neatly support one theory exclusively; there are usually a few conflicting facts. Contrary opinions should be expressed along with the reasons that they are not justified.

To determine the actual cause of failure from several possible mechanisms, I developed the "matrix method of evaluation." This method consists of forming a two-dimensional array to study the relationship between objective and subjective information (see Figure 1.2). Along the vertical axis, all of the relevant facts are listed (objective information). Along the horizontal axis, all of the potential failure mechanisms are listed (subjective information). The intersections are filled in with symbols indicating the degree of support—for example, X for strongly supporting, O for contraindicating, and blank for no relationship.

In this example, brittle fracture appears to be the most strongly supported hypotheses. Note that there is one contradictory finding for each hypothesis, but they have different degrees of significance. This is typical of failure investigations. We usually have some findings that do not neatly fit the prevailing theory, and this reinforces the need for evaluating all possible hypotheses.

Although I have occasionally used the matrix method of evaluation as an actual array, more often I simply use it as a paradigm as I develop narrative discussions about the relationships between the objective and subjective information. The goal is to consider the relationship of each theory with each fact in a systematic method.

Generate Report

In most forensic projects, it is necessary to minimize written documentation because every document is discoverable. Providing a comprehensive written report, however, can save substantial time and money in the discovery process and should be encouraged.

When recording opinions, it is important to list alternative theories—even those that are not supported by the evidence—and explain why. In the adversarial process, it is almost certain that another party will use an omitted fact or theory to impeach the credibility of the report.

Many competent investigators fumble a case at this juncture. Having performed a careful and thorough investigation, they fail to precisely communicate their findings. Perhaps the most common error in this regard is the tendency to confuse *assumptions* with *conclusions*. Although these words may seem too well defined and unrelated to mix up, I have read countless expert reports that do not rigorously segregate these two concepts. To withstand the scrutiny of the adversarial process, it is essential that a forensic engineer be able to enumerate the assumptions made in arriving at his/her conclusions and to recognize that some assumptions are made in arriving at every conclusion.

Manage Exhibits

Forensic projects tend to have enormous files, sometimes with thousands of documents, photographs, and specimens. It is incumbent on the investigating engineer to develop methods of managing all of these exhibits so that they can be preserved, used, and produced to others when needed. In some cases, state or federal rules of evidence may have to be complied with as well.

Objective Information	Brittle Fracture	Fatigue	Stress Corrosion Cracking
Failure occurred suddenly	X		
Failure occurred during cold temperature	X		
Member was under low, static tension	X	O	O
Member was a thick plate	X		
Structure was 10 years old		X	X
Member had moderate corrosion			X
Fracture surface exhibited ductile & brittle zones	O	X	

FIGURE 1.2 Example of Matrix Method
of Evaluation

Prepare for and Provide Expert Testimony

Although expert testimony is not the objective of forensic engineering, it is an occasional consequence of specialized expertise. Generally, a forensic project should be considered complete before a trial (or deposition) begins, so that testimony merely amounts to giving an oral report of findings, conclusions, and recommendations already developed.

Publish a Case History

Finally, publish your findings. If it is not appropriate to use the name of the project, a generic case history can be written. Only through the dissemination of failure information can similar incidents be prevented.

CHAPTER 2. CASE HISTORIES

The case histories presented in this section were first published as monthly columns entitled "Failures," which began running in January 1994 in *The Construction Specifier*. Each column explores a critical detail in construction, taken from a wide variety of disciplines. Case histories are an important part of the engineering literature, because through them vital information is disseminated on what can go wrong in construction. Although they can sometimes cause embarrassment or controversy, they are essential in reducing repeat occurrences.

The column runs on the last page of the magazine. Although it is flattering to have this editorially strong placement, this position severely constrains the length; after headlines, bylines, and a photograph or drawing, there is only room for a few hundred words. Therefore, only the most essential information can be conveyed in each column.

The columns are presented here as they originally appeared, but they are followed by additional discussion. I am fortunate to have the opportunity to augment the brief columns with these additional comments.

Although the columns were generally published monthly, no columns appeared in the following issues: February 1994; December 1994; February 1995; February 1996; January 1997.

The Too-Short Shelf Angle

Building failures are not often catastrophic. Usually they involve parts of facilities that do not perform as intended by the owner, designer, or builder. But the cost of failures—even seemingly minor ones—can be excessive, in terms of both repair and, in commercial and industrial facilities, lost hours.

Beginning this month, I will be presenting critical details—the kind that, handled improperly, can lead to failure. Examples will be taken from a wide variety of disciplines within the construction industry, including structural engineering, masonry, waterproofing, curtain walls, and roofing.

Many of the details will be submitted by members of the Technical Council on Forensic Engineering (TCFE). A technical activity committee of the American Society of Civil Engineers, the TCFE was created in 1982 to develop procedures for reducing the number of building failures. The council disseminates information on failures and their causes, provides guidelines for conducting failure investigations, and encourages research, education, and ethical conduct in the forensic engineering field.[1]

Courtesy the author

By David H. Nicastro

A Crack Case

To kick off this column, I have selected a detail from my own case files (singling out one wasn't easy): the too-short shelf angle.

In multistory construction, the shelf angle, also referred to as a relief or relieving angle, is intended to transfer the weight of masonry back to the structural frame. Without shelf angles, the weight of each floor height of masonry would be carried by the masonry below, causing unacceptably high forces to accumulate. If the steel angle is cut short of the building corners, however, the purpose is defeated: the interlocked masonry bears on two different supports—the steel angle and the adjacent masonry.

In addition to weight, thermal expansion and normal wall movement introduce forces into the masonry that should be carried by the shelf angle. Where the masonry is not fully seated on a shelf angle, these forces are not transferred to the building frame. Therefore, a shear stress is induced at the interface between the forces being transferred into the shelf angle and into the masonry beyond the end of the angle. If the stress is greater than the strength of the masonry, a crack will develop.

The freestanding column of masonry created by such a crack may not be stable. Even with proper brick ties, the masonry beyond the shelf angle's support is isolated from the building frame. Because this type of crack is likely to continue for multiple floors, the end prism of masonry could fall out of the wall.

Proper structural detailing should call for the shelf angles to continuously support the entire perimeter masonry, except for gaps at expansion joints. Other requirements should include

• providing shelf angle support at least two-thirds the width of the brick wythe

• using galvanized or stainless steel to prevent corrosion

• leaving space under the angle to prevent bearing on the masonry below

• using proper flashing.

Of course, each of these could be the subject of an entire article! ◆

Note

1. The TCFE produces the *Journal of Performance of Constructed Facilities*, a quarterly technical journal presenting in-depth case histories of failures and their causes. For subscription information, contact ASCE Marketing Services at (800) 548-ASCE.

DAVID H. NICASTRO, P.E., is a principal engineer in the Houston office of Law Engineering. He specializes in the investigation and remedy of construction problems.

Failure Mechanism: Differential Support

From all of my previous failure investigation cases, I selected the subject for the auspicious first column because the issue is so prevalent. Although masonry construction has been used for thousands of years, in the last few decades much of the technology seems to have been forgotten. As we have moved away from short, stout bearing walls, and moved toward veneer construction, numerous failure mechanisms have been introduced. It is now disappointingly rare to see a modern masonry building that follows the fundamental principles of proper design and construction.

To avoid the differential support condition described here, shelf angles must be fabricated the full length of the prism that they support. We see not only shelf angles that are too short but also many that are too long: they penetrate intersecting planes so that the gravity load of one prism unintentionally bears on another prism. This causes crushing below the shelf angle ends.

One reason for the increase in masonry flaws is that engineers have abdicated any responsibility for brick veneers. The original structural drawings for the building discussed in this case history showed the brick veneer as a phantom line next to the shelf angles. This is typical of current engineering practice; the veneer is not considered an engineered system, so it is not detailed by the structural engineer. Its presence is recognized by the phantom line, but the critical steel/masonry interaction is not contemplated.

Because engineering is the only profession in the construction industry that focuses on preventing failures, it is essential that engineers assume responsibility for the behavior of all building components.

Christine Beall, an eminent architect, wrote a letter to the editor regarding this column indicating that it is sometimes acceptable to construct foundation-supported, multistory masonry walls without shelf angles. The advantages of such a system, when properly designed and constructed, are obvious—it avoids corrosion issues and steel/masonry interaction problems. This column, however, did not mention the type of construction or the building height. Shelf angles are necessary in most applications to keep loads and irreversible moisture growth from accumulating from multiple stories. Given that they are often necessary, it is essential that shelf angles be properly detailed in conjunction with the masonry veneer.

The Deck Drain That Doesn't

This month's critical detail was submitted by Kimball Beasley, P.E., a senior consultant with Wiss, Janney Elstner Associates in Princeton, New Jersey. Mr. Beasley, who is a member of the Technical Council on Forensic Engineering's Committee on Practices to Reduce Failures, specializes in serviceability failures of building components. Beasley's submission is a common waterproofing problem: inadequate drainage.

Trapped Water

Inadequate drainage results not so much from an inability to direct the flow of water as from a failure to extricate it from concealed places once it has been collected. Beasley offers the example of the exterior building terrace or inset roof area with a walking surface of quarry tile pavers set in a mortar bed directly over a single-ply waterproofing membrane, surface drainage being provided by deck drains (see figure).

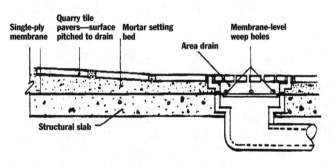

Single-ply membrane / Quarry tile pavers—surface pitched to drain / Mortar setting bed / Area drain / Membrane-level weep holes / Structural slab

By David H. Nicastro

Many buildings with exterior terraces have surface drainage only, which is clearly a misunderstanding of basic waterproofing design, i.e., that the membrane level must be sloped and drained in addition to the visible top level. Even though *most* of the water impinging on a terrace will run off the surface into deck drains, some water inevitably filters down to the membrane. Unfortunately, many designs fail to differentiate water shedding from waterproofing.

Including dual-level drains with weep holes in the collar to admit water from the membrane level is not a complete solution. Without free flow along the membrane, the water is trapped. Since the membrane is impervious and the tile above it prevents rapid evaporation, the water has to percolate through the mortar setting bed. Beasley suggests that this problem is more pronounced with new single-ply membranes than with older coal tar types because the latter allow some vapor transmission.

Water trapped below the walking surface can cause serious problems. Over time, water-soaked mortar deteriorates. Also, a continuously wet membrane is more likely to leak than one that drains freely. Further, tar-like constituents of some bituminous membranes can boil or float up with the trapped water, making unsightly stains on the deck surface. And when the water does find an exit, the leached salts and calcium carbonate from the setting materials can build up, causing efflorescence.

The solution to all of these problems is simply to install a free-draining fabric or drainage course over the membrane. Together with a dual-level drain and a sloped deck, this technique will eliminate the water problems associated with terrace decks. Alternatively, a pedestal system with open paver joints may be used to allow free drainage at the membrane level.

Short of a total tear-off and replacement, the best that can be done to fix existing terraces is to minimize the amount of water that filters through the surface. Careful maintenance of joints, cracks, and sealers will limit the amount of trapped water. Once deterioration has begun, however, surface sealers usually accelerate failure of the pavers and mortar setting bed.

While limiting the water migration through the surface is like "sticking your finger in the dike," a simple additional measure will often solve the problem: by installing a perimeter trench drain, free drainage can be provided at the membrane level in critical areas without a total tear-off. ◆

DAVID H. NICASTRO, P.E., is a principal engineer in the Houston office of Law Engineering. He specializes in the investigation and remedy of construction problems.

Failure Mechanism: Not Applicable

This column explored one of the most common detailing errors in commercial construction: a "bathtub" waterproofing system. This type of design is contrary to fundamental waterproofing principles. An effective waterproofing design must contain three components:

1. *Collection* is necessary to direct water to predetermined locations in a controlled manner, without allowing uncontrolled migration.
2. *Drainage* is necessary to provide an adequate means of escape for the collected water.
3. A *back-up system* or "second line of defense" is necessary so that no single component is critical to the overall system integrity.

In my experience, most water infiltration occurs simply due to a lack of drainage. If water is permitted to drain freely away from building materials, they do not need to be installed perfectly; this is flaw-tolerant design. But when water collects on or in a system, such as the deck discussed in this column, then the materials and workmanship must be without defects or the water will certainly find them. Such flaw-intolerant design imposes unrealistic expectations on the construction industry.

At the end of the column, a suggestion is made for a remedy that does not require total tear off and replacement. In my experience, simply adding drainage channels at the perimeter of a deck will often suffice. This approach is feasible if the leaks are concentrated at the perimeter and not dispersed throughout the field of the deck. This indicates that the deck itself is reasonably watertight and serves as a collection system to bring the infiltrating water to the low spots or vertical interruptions. If the water is then allowed to drain off at these collection points (such as by new subsurface drains), the leakage should be abated.

Of Stress and Stucco

Cracks in stucco are so common that owners become inured to them, often accepting as normal what are really inexcusable failures caused by poor detailing. In addition to being unsightly, cracks can lead to more serious problems, such as water infiltration, spalling, or loss of structural integrity.

Balcony Connections

According to Kenneth Simons, chairman of the American Society of Civil Engineers' Committee on Practice to Reduce Failures, one of the most common stucco systems encountered today consists of portland cement 20 mm (¾ in.) thick applied directly to expanded metal lath over asphalt-impregnated building paper and gypsum wallboard 16 mm (⅝ in.) thick. The wallboard is fastened to light-gauge metal studs, and a colored acrylic finish coat 3 mm (⅛ in.) thick is applied to the portland cement plaster.[1]

On one project Simons knows of, this type of system was used to create uninsulated, hollow barriers between private, prefabricated steel balconies fastened through the stucco into wood blocking. The balconies were separated from the walls by neoprene washers (see figure).

The problem with this detail was that the washers were only effective for compression,

> The stucco's relatively low capacity in tension resulted in cracking problems.
>
> By David H. Nicastro

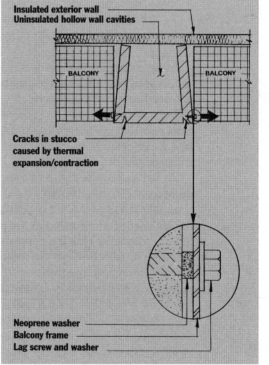

Insulated exterior wall
Uninsulated hollow wall cavities

BALCONY   BALCONY

Cracks in stucco caused by thermal expansion/contraction

Neoprene washer
Balcony frame
Lag screw and washer

i.e., expansion of the rail against the wall. In tension, the stresses from the contraction of the balcony rails and the cavity walls were transferred through the lag screws into the stucco. If the walls had been able to withstand the stress induced by the accumulated thermal contraction of the rails and cavity walls, there would have been no problem. But because the stucco was brittle and had a relatively low capacity in tension, cracks developed.

Simons says providing adequate clearance between the railings and the stucco and screws would have reduced the cracking. Also, a different type of connection could have been used; since the support needed for the balconies and rails is only perpendicular to the axis of the rails, dowels could have been used in oversized holes. ◆

Note

1. The United States Gypsum Association's *Gypsum Construction Handbook* specifies that stucco wall panels should be relatively square, with no dimensions exceeding 5.5 m (18 ft) without a joint. The handbook also states that control joints should be installed for surface areas of approximately 14 m² (150 ft²), at penetrations, and at typical points of stress concentration. These conditions were met in Simon's example, but cracks still developed.

DAVID H. NICASTRO, P.E., is a principal engineer in the Houston office of Law Engineering. He specializes in the investigation and remedy of construction problems.

Failure Mechanisms: Thermal Expansion; Cracking

The footnote to this column raises an interesting issue: even when conventional design guidelines and building codes are thoroughly followed, some failures still develop. In this case, the particular design error was never contemplated by the industry standards. We are often asked by owners to investigate whether unsatisfactory construction is in compliance with applicable codes, and they are surprised to learn that it is.

It is also interesting to note how many construction failures are related to thermal stress. It would seem that this fundamental principle of physics—that things expand on heating and contract on cooling—would be well understood. We continue to see designs, however, that do not accommodate the actual thermal behavior of a component or system.

Thermodynamic effects on buildings are very complex. A component expands and contracts according to its own temperature, which is a function not only of the ambient temperature but its reflectivity (solar gain), porosity (evaporative cooling), and more subtle phenomena such as black-body radiation. Therefore, the temperature range that a component experiences may be substantially larger than that of the ambient air temperature on a seasonal basis—perhaps twice as large.

Furthermore, the published coefficients of thermal expansion may not be accurate. Although metals tend to have well-established coefficients, concrete and other variable products have variable properties; an upper-bound value must be assumed in design. Even then the coefficients may not be constant: for some materials, the thermal expansion is nonlinear over the service temperature range, and the coefficient of thermal expansion of some porous materials increases substantially when they are saturated with water—more than 50% in some cases.

Heating and cooling cause a material to expand or contract unless it is restrained (like contraction of the balcony handrail in this case). Then the strain that would have occurred is converted into stress, according to the principles of superposition and the material's modulus of elasticity. The simplistic calculation of thermal expansion—multiplying change in temperature by the material's coefficient of thermal expansion—yields the ideal change in length for a fully unrestrained specimen. Any actual installed component would behave somewhere between fully restrained (all stress) and fully unrestrained (all strain). Analyzing the actual degree of restraint can be more difficult than determining the ideal thermal expansion of the component.

Because of nonlinear thermal expansion, variable thermal coefficients, and the difficulty in calculating restraint and the actual service temperature range, we can expect many more thermally induced failures.

Skylight Hang-Ups

One Monday morning in 1987, maintenance personnel at a major art museum in New York discovered that a portion of the glass ceiling over a public exhibit space had fallen 15 m (50 ft) to the floor. Fortunately, the catastrophe occurred before the museum had opened, and no one was hurt.

The ceiling, which had been installed about 10 years earlier, consisted of wire glass panels supported by an aluminum framing system hung from the exposed steel structure with rods 5 mm (³⁄₁₆ in.) in diameter.

According to David B. Peraza, an associate of New York-based Thornton-Tomasetti Engineers who specializes in the investigation of structural building failures, the ceiling was unusual in several respects. First, the roof of the exhibit space consisted entirely of skylights. This caused wide temperature fluctuations in the ceiling space. Second, some of the hanger rods were extremely short. While most were 0.6 m (2 ft) long, those along the perimeter, where the ceiling sloped up sharply, were only 38 mm (1.5 in.) long (see figure).

Bending forces due to thermal movement led to hanger failure.

By David H. Nicastro

Peraza's examination revealed that six of the short hanger rods had fractured, and that several of the fractured rods were kinked—one as much as 14 degrees. The kinks suggested that the ceiling had been moving horizontally.

Metallurgical examination indicated that the fractures were initiated by fatigue; specifically, by extremely high stress over relatively few cycles. Adjacent rods were tested and found to have a safety factor of 15 against yielding in tension. Peraza hypothesized that bending forces were caused by differential thermal movement, and, in fact, calculations showed that a temperature change of only 1.7 °C (3 °F) was sufficient to cause yielding in the short rods.

Very high bending stresses were induced by differential thermal expansion and contraction

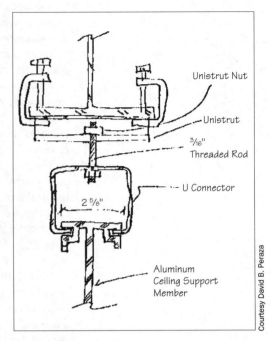

Unistrut Nut

Unistrut

³⁄₁₆" Threaded Rod

U Connector

2 ⁵⁄₈"

Aluminum Ceiling Support Member

Courtesy David B. Peraza

of the ceiling frame relative to the steel roof framing: aluminum has a coefficient of thermal expansion about twice that of steel, so when the sun shone through the skylight, the ceiling frame expanded horizontally twice as much the roof frame.

The long rods could accommodate the thermal movement by tilting slightly. The perimeter rods, however, could not tolerate the displacement as easily. Due to their shortness and to the partial rotational restraint at the top and bottom afforded by the nuts, the rods were forced to bend. Much as a paper clip can be broken by bending it back and forth, the movement eventually fractured the rods.

A new detail developed for the short hanger connections included thicker rods and teflon plates to allow horizontal movement. No remedial work was deemed necessary for the long hanger connections. ◆

DAVID H. NICASTRO, P.E., is a principal engineer in the Houston office of Law Engineering.

Failure Mechanisms: Thermal Expansion; Fatigue; Fracture

This case, like the previous one, involved a failure caused in part by thermal expansion and contraction. But the proximate cause of the failure in this case was fatigue.

Scraping marks observed at some of the connections of the short hangers indicated that, at first, they had slid to accommodate the thermally-induced movement. Eventually, the sharp edges of the square nuts dug into the kendorf channels, preventing further sliding. Then they attempted to tilt, but were forced to bend.

As with the masonry veneer case history (page 8, "The Too-Short Shelf Angle"), the structural engineer was not involved with the design of the system. It is customary for an architect to design skylights and suspended ceilings without an engineer's involvement, which is acceptable for ordinary systems. But specialized construction or deviations from conventional design practices should be reviewed by a structural engineer.

Precast Improvisations

Concrete beams are typically reinforced with steel bars for flexural strength. Equally important—but often ignored—is the need for shear reinforcement, or stirrups. Improperly spaced stirrups are a primary cause of structural distress in concrete beam construction, sometimes leading to collapse.

Edward R. Fisk, vice president of Construction Services for Wilsey &

Unwarranted "improvements" during construction caused beam failure.

By David H. Nicastro

Ham in Foster City, California, reported the following incident regarding improperly spaced reinforcement.

Playing It By Ear

At a precast concrete fabricating plant, Mr. Fisk, who is a civil and structural engineer and general contractor, observed a laborer carefully arranging stirrups in a beam form so that they were spaced equally—even though the engineer's drawings called for unequal spacing. The laborer even provided an "extra one."

After erection, a progressive failure of the beams occurred.

The structural engineer intended for the stirrups to be placed over the support area of the beam (see figure). This placement is essential to

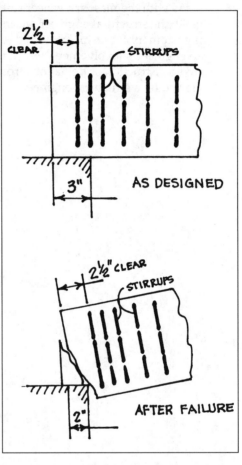

shear reinforcement, as the plane of highest shear is at the face of the support. Some reinforcement must be on each side of this plane.

The laborer, however, attempted to provide 75 mm (3 in.) of concrete cover for the first stirrup by moving it away from the end of the beam. Compounding the problem was the engineer's miscalculation of the amount the member would be shortened during prestressing. Shortening resulted in 50 mm (2 in.) of end bearing, with the first stirrup 25 mm (1 in.) farther out. During the failure investigation, no stirrups were found in the exposed fractured area.

Construction laborers are generally unfamiliar with the engineering principles involved in reinforced concrete design. In this case, the laborer was following what he thought was common sense: equal spacing, good cover, and plenty of steel (the extra stirrup). But the failure resulted simply from having no reinforcement in the critical region.

Two things are needed to prevent such failures: 1) monitoring (design follow-through) and 2) flaw-tolerant design. The engineer's underestimate of beam shortening would not by itself have resulted in failure, but it contributed by making stirrup placement overly critical. A larger bearing area would have accommodated the very common construction error of misplacing the shear reinforcement. ◆

DAVID H. NICASTRO, P.E., is a principal engineer in the Houston office of Law Engineering.

Failure Mechanisms: Overstress; Diagonal Cracking;
Diagonal Tension Failure

This is another example of flaw-intolerant design. Because the fabricator deviated from the engineer's drawings, some minor deviations should be expected. Overly critical designs are failure prone and unnecessary in building construction. Construction, at its best, is imperfect, and engineering designs should take this into account, to a certain degree.

Even with noncritical design, monitoring during fabrication and construction is prudent to ensure that the contractor's personnel correctly understand the design intent. I am certain that one of "Murphy's laws" must be "anything that can be misinterpreted will be."

12,000 Points of Water Entry

The isometric cross-section shown here is the as-built configuration of a high-rise office building curtain wall completed in 1984. This appears to be an elegant design satisfying the key functions of all exterior building envelopes: aesthetics, resistance to water and air infiltration, thermal isolation, and structural performance. One small detail was overlooked, however.

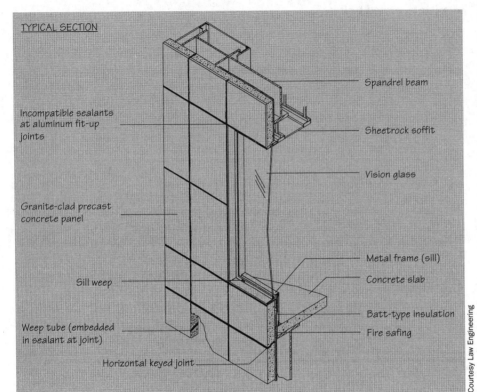

TYPICAL SECTION

Spandrel beam
Incompatible sealants at aluminum fit-up joints
Sheetrock soffit
Vision glass
Granite-clad precast concrete panel
Metal frame (sill)
Sill weep
Concrete slab
Batt-type insulation
Weep tube (embedded in sealant at joint)
Fire safing
Horizontal keyed joint

Courtesy Law Engineering

By David H. Nicastro

Sealant Mortality

A slice through almost any section of the wall would cut through at least two materials. This redundancy provides a backup defense against water infiltration. Furthermore, the various cavities created by the assembly of different components and hollow aluminum extrusions are drained to the exterior through weep holes. Thus, the three key elements of good waterproofing design—drainage, collection, and backup—are adequately provided in most cross sections. Except at the window perimeter joints.

The only thing between the outside and the inside of the building at the interface of the windows and panels is a bead of elastomeric sealant. Even when correctly specified and installed, however, elastomeric sealants are unlikely to survive for the life of a building. At the time the sealants are replaced, failures typically are well advanced, and the building has already experienced leakage problems.

Incompatibility

Although not a good detail, in this instance the single sealant bead around the windows should not have been a problem until the material reached the end of its design life. But a compounding problem caused premature failure: the perimeter joint sealant material (a polyurethane) was incompatible with the internal window extrusion joinery sealant (a silicone). The perimeter sealant disbonded and curled away from the joinery sealant where the two were in contact at the mitered corners of each window. Although only a small defect, the cumulative effect was equivalent to poking 12,000 holes in the building envelope.

To remedy the problem, the window perimeter sealant was cut out and replaced. Special care was used in cleaning the window frames of any sealant residue, which had originally exuded from the mitered corners, causing the problem in the first place. ◆

DAVID H. NICASTRO, P.E., is a principal engineer in the Houston office of Law Engineering. Mr. Nicastro specializes in the investigation and remedy of construction problems.

Failure Mechanism: Reaction

In this investigation, it was interesting to find an inconsistency in the curtain wall design concept. Whereas all of the other cross sections had redundant waterproofing designs, the sealant joints around the perimeter of the windows did not. This is an example of the over-reliance designers place on elastomeric sealants and flaw-intolerant design. Sealants were used as though they were designed as void fillers to make up for construction tolerances, rather than as primary seals in a redundant system including drainage.

Elastomeric sealants have been instrumental in the development of modern architecture, making curtain walls and veneer construction possible. Their durability, however, is questionable. Organic sealants (such as the urethane used in this case) deteriorate over time, requiring replacement. Inorganic sealant (silicone) does not break down under the influence of weathering, but the durability of the adhesion of the cured rubber to the substrate is a function of workmanship, which again must be considered imperfect.

Because of the inevitability of some degree of failure, no exterior sealant joint should be critical to maintaining a waterproof closure. Whether due to poor workmanship, inappropriate materials, poor design, or normal deterioration, some maintenance or replacement of exterior sealants must be expected within the service life of any building. At what point maintenance is performed is determined by several variables.

- A flaw-tolerant, redundant design can accommodate some defects and deterioration without resulting in leakage or other problems, so remedial work can be deferred.
- Some building owners (especially institutional) cannot respond rapidly to evidence of deterioration, no matter how profound, so it may progress for extended periods even after discovery.

Buildings That Lose Their Marbles

Early one winter morning in 1989, a piece of marble cladding fell off a high-rise office building in Houston, Texas, shattering on the sidewalk. No one was hurt in the incident, but an

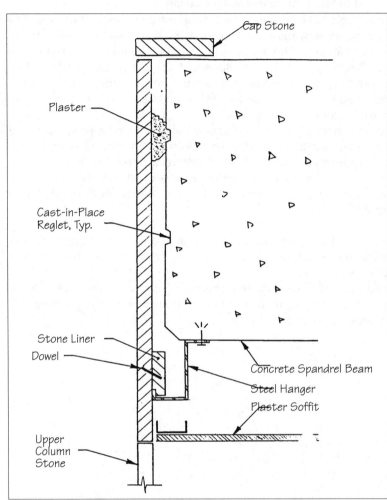

Cap Stone

Plaster

Cast-in-Place Reglet, Typ.

Stone Liner

Dowel

Concrete Spandrel Beam

Steel Hanger

Plaster Soffit

Upper Column Stone

Courtesy the author

By David H. Nicastro

epidemic of similar failures throughout the country has since resulted not only in injury to passersby but in staggering economic losses—replacing exterior cladding can cost more than an entire building cost in the first place.

More than one failure mechanism was at work in Houston, but warping of the thin marble panels was the predominant problem. There have been many anecdotal reports of marble cladding "following the sun" and "warping like plywood" when exposed to the elements. The leading scientific explanation is "thermal hysteresis."

Sugar...

Thermal hysteresis is a permanent, progressive volume change. In the thin, calcite-rich marble typically used in American cladding, differential heating causes the outer face of the panel to expand, after which the calcite crystals interlock, preventing the panel from returning to its original shape.[1] (Marble does not experience this problem when it is used in solid blocks, as it was 30 years ago.)

Because of the distortion, the marble becomes "sugared" and loses strength. The distortion also increases the loads on lateral anchors. The latter phenomenon alone would condemn many facades.

...and Dust

The other major form of deterioration contributing to the Houston failure involved the marble panels' concealed anchors.

As is common practice, the thin stone panels were supported by "liners"—small pieces of stone adhered to the back of the face stone to bear on the shelf angle. The adhesive used for the Houston office tower, however, had turned to dust. With their main gravity support gone, panels were left hanging by their lateral ties.

Apparently, the panel that fell, already weakened by thermal hysteresis, cracked in an intense cold snap, leaving a piece of stone that was neither supported by a gravity anchor nor restrained by a lateral tie. It was this piece that fell to the sidewalk. ◆

Note

1. Bernie Erlin, "When Is a Bow Not a Bow?" *Stone World* (February 1989): 86.

DAVID H. NICASTRO, P.E., is a principal engineer in the Houston office of Law Engineering. He specializes in the investigation and remediation of construction problems.

Failure Mechanisms: Thermal Hysteresis; Aging; Sugaring; Fracture; Thermal Shock

Numerous failure mechanisms were at work in this case. The event that precipitated our investigation was that one-half of one stone fell off of the building. During the course of our study, we quickly determined that the remainder of the stones was inadequately supported—for entirely different reasons.

We concluded that the stone that fell had first fractured horizontally, probably due to thermal shock; the temperature had fallen precipitously just before the failure. The upper portion of the stone rotated away from the wall, hinged about the crack, and fell free because the lateral anchors failed. The stone had lost strength over time (as is common with marble), reducing the capacity of the anchors. Besides, the lateral anchors were not designed to carry the gravity load of a free-swinging, broken stone.

Only because of this event did we discover that the epoxy adhesive used to secure the liners had deteriorated over the 30-yr service life. After the epoxy turned to dust, the dowels created concentrated loads, thereby fracturing the liners. This caused enormous displacements of stones—measuring up to 3 in.—but they were still hanging by their lateral anchors.

The large displacements had been observed by tenants in neighboring buildings prior to the falling incident, and the owner had already contacted an engineering firm about studying the distress. As ironic as it may seem that the failure occurred so soon after the first distress was observed, this sequence is disturbingly common. One of the most frequent events generating new projects for my engineering firm is exterior components (especially stone and masonry) falling off of high-rise buildings—usually long after severe distress was present. It is tragic that in some cases (although, fortunately, not this particular one) the falling pieces hit people and caused severe personal injury. Some jurisdictions have enacted local ordinances requiring frequent inspection of building facades to locate proactively such distress and to prevent these incidents.

The stone's loss of strength and warping were caused by thermal hysteresis, a process usually associated with marble, which also occurs with other materials. There is an epidemic of marble veneer facades suffering from thermal hysteresis throughout the United States. The process takes years to weaken the stone to the point of causing failure, and those failures are beginning to be actualized on many buildings built since 1960—the era when thin marble veneer (<2 in. thick) was technologically possible and popular.

The combination of these last two points—that there is an epidemic of incipient marble failures and that many owners do not take action until stone actually falls off—should give any pedestrian pause for thought.

Cracking in GFRC

Glass fiber-reinforced concrete (GFRC), introduced in the United Kingdom in the 1960s, gained popularity in the United States during the 1970s. A few manufacturers still produce GFRC panels in this country.

Early on, studies of these composite panels showed that they lost tensile strength and ductility over time. This weakening process, known as "aging," can lead to relatively rapid cracking. For example, in a 1986 investigation of a re-

Early aging caused loss of strength and ductility in composite panels.

By David H. Nicastro

cently completed San Antonio office building, I observed extensive cracking in the building's GFRC spandrel panels. The cracks, which occurred in the architectural reveals, had been camouflaged with brown paint (see photo).

Courtesy the author

Factors in Aging

GFRC panels are usually made by spraying a cement/sand matrix and short lengths of alkali-resistant glass fiber into molds in a precasting yard. The fibers are chopped and introduced into the mixture at the spray nozzle. A less common fabrication method is to mix the fibers in the mortar, then cast, mold, or extrude the final product.

Two theories have been propounded to explain GFRC aging. One is that alkali attack on the glass results in reduced fiber size. The other is that hydration products penetrate and fill up the interstitial spaces of the fiber bundles, decreasing the ability of individual fibers to slip relative to the cement matrix.

Because of the inherent loss of strength and ductility, the fiber manufacturers recommended limiting the use of GFRC to applications primarily in compression. Unfortunately, this recommendation was not widely recognized. Cladding panels, for instance, experience tensile forces as they transfer wind loads from the face to connections. Shrinkage and thermal stress in panels can also exceed the panels' aged strength, causing more cracking mechanisms.

Cracking is further exacerbated by panel thinness: typical composite panels are only 20 mm to 40 mm (¾ in. to 1½ in.) thick, which means they act more like a membrane than conventional precast concrete panels. Some buildings clad with these thin panels have required entire facade replacement.

A Simple Solution

Because the cracks on the building in San Antonio were limited to the reveals, repair was straightforward: the reveals were filled with silicone sealant over a bond-breaker tape.

Incidentally, we never discovered who painted the brown stripes over the cracks. Some mysteries are never solved by failure investigation. ◆

DAVID H. NICASTRO is founder and president of Engineering Diagnostics in Houston, Texas. Mr. Nicastro specializes in the investigation and remedy of construction problems and the resolution of related disputes.

Failure Mechanisms: Aging; Crystal Packing; Necking; Etching; Shrinkage; Embrittlement

The use of GFRC has diminished substantially since the 1980s, but it is still manufactured. This column prompted a response from a GFRC fabricator who indicated that aging is insignificant if the panels are properly designed for movement. My research, however, indicates that the loss of ductility over a period of 10 yr renders a typical panel effectively as brittle as unreinforced concrete. The notorious cracking prevalent in GFRC projects can be primarily explained by aging, which leaves a thin, brittle shell of concrete on flexible supports.

Moisture is involved in both of the recognized embrittlement mechanisms: necking and crystal packing. Therefore, cracking can theoretically be prevented by prohibiting moisture migration through the panels. Attempts at arresting cracking by remedially applying water sealers, however, have met with mixed results; some buildings have no new cracking, whereas others continue to crack unabated.

Part of the cracking in these panels was attributed to bridging the vertical panel joints by the aluminum window sill and head extrusions. Although the ribbon windows and the panels both had expansion joints at the same intervals, they did not line up. Because the window frames were bolted to the panel frames, the two systems effectively reinforced each other. We investigated the particular connection detail in attempting to discover whose responsibility it was. We were surprised to find that the window and panel shop drawings both pointed to the connection with the note "by others." Apparently, no one detailed the connection.

Skin Cancer

When I first saw the Park Square condominiums in Houston, Texas, in 1988, the only noticeable distress in the exterior cladding consisted of a few cracks and spalls.

The high-rise luxury condominium project, built in the early 1980s, was clad with Granostruct panels. The prefabricated panels consisted of cement-asbestos board (CAB) sheathing fastened to welded steel stud frames. The CAB was coated with marble chips broadcast onto a resin binder. The panels looked somewhat like exposed-aggregate

A fine network of cracks hinted at widespread cladding disease.

By David H. Nicastro

precast concrete panels, but they were significantly lighter in weight.

Unfortunately, the minor visible distress was the surface manifestation of "cancer of the building skin." Because the finish was not reinforced with mesh across CAB joints, cracks had occurred readily at the joints when the cement-rich panels moved and shrank. When the building was viewed up close, cracking was found to be extensive.

Upon removal of a portion of the cladding, it was discovered that water infiltration through the myriad small cracks had corroded the concealed steel studs almost to the point that they were no longer serviceable. Full-scale, in-place load tests showed that the studs could continue to be used if the water infiltration was arrested quickly.

Courtesy David H. Nicastro

Remedy

Four levels of potential scope are usually considered when a course of remedial action is selected:

1. Repair only current distress, and monitor for future distress.

2. Repair current distress, and fix all locations likely to suffer similar distress.

3. Add a new system to the existing one, making the failed system obsolete but leaving it in place.

4. Remove the failed system and replace it with a new one.

Obviously, cost and time increase dramatically with each higher level of remediation; therefore the lowest practicable level is usually chosen.

Because of the steel corrosion at Park Square, waiting to begin remedial work almost certainly would have required jumping to level 4, which was unacceptable since removal of the studs would have entailed "daylighting" the condominium units. Levels 1 and 2 were not feasible, simply because the rough aggregate surface made adhering a sealant bead to the panels impossible; attempted sealant remedies failed because the blade bounced off the stones during tooling.

Ultimately, level 3 was chosen, and the entire building was reclad with a new exterior insulation and finish system, leaving the existing exterior in place. ◆

DAVID H. NICASTRO is founder and president of Engineering Diagnostics in Houston, Texas. Mr. Nicastro specializes in the investigation and remedy of construction problems and the resolution of related disputes.

Failure Mechanisms: Embrittlement; Shrinkage

This column describes the failure of one brand of aggregate-surfaced panels, but there were similar panels made in the 1980s by many companies. Like GFRC, these panels were a substitute for precast concrete, but they lacked the durability of reinforced concrete. Numerous similar failures occurred, and many buildings were reskinned to alleviate the distress.

According to the manufacturer, earlier panels of this type had performed well, but then the cement–asbestos boards suddenly became much more brittle. The manufacturer theorized that this change in behavior was because the later boards were made with shorter asbestos fibers, which reduced the panels' ductility. Spontaneous cracking destroyed the panels, and the failures subsequently bankrupted the company.

In this investigation, it was very difficult to find the cracks in the panels. They were hard to see against the bright white background, they meandered around aggregate, and they closed tight in sunlight; cracks that admitted water during testing in shade did not leak when in direct sunlight. We found that the longer we looked at any one panel, the more cracks we could find.

The severity of the corrosion of the steel studs necessitated a comprehensive remedy. In addition to leakage through the cracks, the sealant joints contributed significantly to water infiltration. (See "Little Changes, Big Problems", page 32, for more information.)

Wrong-Headed Remediation

Perhaps the most frustrating building failures are those resulting from inappropriate remedial work. Consider the example submitted to me by Kimball Beasley, P.E., a senior consultant with Wiss, Janney, Elstner Associates in Princeton, New Jersey. Beasley specializes in evaluating serviceability failures of building components.

As the photo shows, a masonry shelf angle was sealed over during remedial waterproofing, though the conventional wisdom of masonry construction is that water will drain

Inexperienced contractors and designers are dabling in building repair.

By David H. Nicastro

out at shelf angles, whether the wall is solid or cavity-type. A sealant bead placed over the toe of the angle without weep holes or tubes can trap water against the steel inside the wall, leading to corrosion, leakage, or masonry deterioration due to formation of ice crystals.

In Beasley's example, loss of cross-sectional area made replacing the shelf angle necessary. (Above windows, where sealant is often inappropriately placed over steel lintels, it is often possible to cut out the offending sealant, grind off the scale, and dry-pack the void with mortar, leaving weep spaces.)

A Panacea

Beasley's example reflects a prevalent misconception—i.e., that elastomeric sealants can be used indiscriminately as a panacea for leakage problems. Unless the causes of water infiltration and the entire path of migration from outside to

Courtesy Kimball Beasley

inside are known, however, adding sealant to non-original details may not only be ineffective, it may magnify problems.

In one case, I found that sealant had been injected into the glazing channel of a leaking ribbon window system on an entire eight-story office building, when what the system actually needed was bridge flashing between adjacent sill extrusions. All the glass had to be removed and the sealant cut out before internal repairs could be made to the windows. The mistake cost the owner $300,000.

Such incidents are on the rise. Due to a poor market for new construction, contractors and designers are moving into the remediation marketplace without the necessary training and experience. Owners should therefore be wary of overly simple solutions (and supposed bargains).

As a wise man once said about "expensive" experts, "When you have to hire one to undo the work of an amateur, they don't seem so expensive after all." ◆

DAVID H. NICASTRO is founder and president of Engineering Diagnostics in Houston, Texas. Mr. Nicastro specializes in the investigation and remedy of construction problems and the resolution of related disputes.

Failure Mechanisms: Cracking; Corrosion;
Ice Crystal Formation; Necking

This column discusses a subject that is even more shameful than poor original construction: poor *remedial* construction. Imagine the anger and frustration of an owner who has a major building problem worsened by inappropriate remedial work. This scenario is disturbingly common, partly because remedial design is not taught in engineering schools: it should be. There is an increasing need for specialists in rehabilitation of older structures.

The photograph illustrates cracking and crushing below the shelf angle. Inadequate horizontal expansion joints under shelf angles are a primary cause of distress in masonry. In this case, the gravity forces transmitted through the shelf angle into the masonry below were exacerbated by corrosion and ice crystal formation, both of which result in internal pressure. The advanced corrosion, caused by trapping water behind the inappropriate sealant, also resulted in the loss of steel section, jeopardizing the structural integrity of the masonry.

Museum Mask

The seaport museum shown in the photographs to the right was recently built on the Gulf Coast. Although local codes required structures to resist winds of 225 k/h (140 mph), the exterior brick masonry walls could not even withstand ordinary maximum design wind loads, as we learned when we were called in shortly after construction to investigate efflorescence, cracking, and water infiltration problems. During storms, water penetrated the cracks in the facade and ran down the interior face of the walls, saturating the carpet.

Structural flaws warranted a clever coverup.

by David H. Nicastro and Nar Sripadanna

We used several techniques to evaluate the structure, including computerized modeling of the wind-induced wall stresses and testing of actual material properties. This evaluation revealed conditions that led us to conclude a structural upgrade would be necessary: masonry units and mortar were not well bonded; steel reinforcing bars were undersized; and reinforcing was discontinuous.

Incognito Repair

We performed feasibility studies of different remedial techniques, considering implementation time, cost, and the sensitivity of the museum interior, which contained a full-wall movie screen in addition to historical artifacts. We concluded that a new thin-brick veneer would have been too expensive and that full-thickness masonry would have been too heavy for the existing foundation.

Instead, we recommended a scheme that consisted of installing structural steel tubes horizontally on the existing walls, then attaching corrugated concrete panels 6 mm (¼ in.) thick using stainless steel masonry anchors. The resulting thickness of the wall section required extending the eaves and window trim.

As the photographs show, the remedial construction dramatically changed the appearance of the building. Whether the change was an aesthetic improvement is a matter of taste. But we were pleased that such a major repair could be made without creating the appearance of a "fix-it job." In fact, the new exterior fits in well with the industrial look of surrounding buildings. We believe most museum visitors will never know the facade they see is a "mask" for one that failed. ◆

DAVID H. NICASTRO, P.E., is president of Engineering Diagnostics in Houston, Texas. **NAR SRIPADANNA, P.E.,** is a senior engineer in Law Engineering's Houston office.

Failure Mechanisms: Cracking; Overstress;
Shrinkage; Brick Suction

As so often happens, we were first called to investigate this building because of water infiltration, but then discovered substantial structural problems. Leakage is often the first symptom an owner notices, but it may indicate that the structure has cracked or moved significantly.

It should not have been surprising that leakage occurred in this building because it was constructed with single-wythe brick masonry exterior walls. The design included no cavity, flashing, or weep holes. The only barriers against water infiltration were brick and mortar. This is an egregious example of flaw-intolerant design: masonry should be expected to be permeable.

This case typified another common scenario in forensic engineering: the more one looks, the more problems one finds. I do not mean to imply that any structure so scrutinized would yield a myriad of substantial problems—only that buildings with shoddy workmanship tend to have problems in multiple systems.

After escalating from water infiltration to masonry wall cracking, we next discovered that the walls were not designed appropriately for the severe wind conditions on the Gulf Coast. The walls were laterally supported in the vertical direction by the foundation at the bottom, the roof at the top, and one or two raised floors. In the horizontal direction, the walls were supported by reinforced masonry pilasters. The design capacity was inadequate in each direction; we even tried an unconventional two-way analysis and still found the capacity insufficient.

Furthermore, the pilaster reinforcement was missing or ineffective in many core holes. In some places, the bars were in place but useless, because the cells were not grouted; the steel was not centered in some walls; the reinforcement was not properly lapped; and corrosion had reduced the section properties of some reinforcement.

The mortar was not well bonded to the bricks. There were visible separations at the brick/mortar interfaces, which were confirmed by microscopic analysis to have been caused by initial workmanship, as opposed to stress-related cracks that occurred later. This condition weakened the flexural strength of the walls.

Next we discovered that the first-floor slab was not thickened under one stub column, which carried second-floor loads down to the first floor. This situation made a punching shear failure possible. Also, the second-floor flitch beams were undersized. Structural remedies were implemented for both of these conditions, in addition to the exterior walls.

Overcladding the walls changed the exterior building dimensions, so the remedial design had to include new window and door penetration treatments and roof eave extensions. The windows shown in the upper windows were false, so they were simply overclad to economize. The museum remained open to the public throughout construction. The work was entirely performed from the exterior; no valuable exhibit space was put out of service.

Unstable Shoring

Soil-structure interaction is one of the most common sources of construction failures. Processes such as shrinking and swelling of clays and settlement of all types of soil can wreak havoc with foundations. Such distress usually does not occur until well into the service life of a structure.

Vulnerability

The photo shows a model of a shoring tower that collapsed in 1982, killing 30 people. Such shoring systems are used to support a variety of temporary construction loads, including highway bridge formwork and plastic concrete loads. The systems generally consist of four-legged scaffolds, with each leg supported on its own footing pad. The legs are cross-braced with relatively light horizontal or diagonal members. Such scaffolding is vulnerable to sudden catastrophic failure under load.

Soil-structure interaction should be considered even for scaffolding.

By David H. Nicastro and Merle E. Brander

In this case, the bridge's main towers were well founded beneath the fill on a suitable soil with predictable settlement behavior. The behavior of the fill, however, could not be predicted, creating the possibility—tragically realized—of settlement varying from one leg to another.

If a single leg of this type of shoring system settles, the load on that leg is transferred to the diagonal bracing, which is typically inadequate for the additional forces. If the bracing fails, one or more of the loaded legs can buckle. Even if the bracing doesn't fail, the remaining legs may buckle under the additional load previously supported by the unstable leg.

Computerized structural analysis of the failed tower indicated that as little as 5 mm (3/16 in.) of differential settlement could cause a collapse. Investigators concluded that independent footings were therefore unsuitable for the application.

DAVID H. NICASTRO, P.E., is president of Engineering Diagnostics, Inc., in Houston, Texas. **MERLE E. BRANDER, P.E.,** is president of Brander Construction Technology, Inc., in Green Bay, Wisconsin. Mr. Brander is a member of the ASCE Committee on the Dissemination of Failure Information.

Courtesy Merle E. Brander

Failure Mechanisms: Differential Settlement;
Overstress; Buckling

This failure occurred during construction. Although most of the case histories presented in these columns are taken from the service phase of a building's life, most failures occur during construction—from 50–70%, according to some surveys.

The construction phase is the breaking-in period for a structure: failure mechanisms that are not time dependent may develop quickly during construction. In addition, there are many factors that contribute to the vulnerability of structures during erection. A few of these factors follow:

- Structures are not stable until all members are connected.
- Very heavy loads and equipment are being moved constantly.
- Material properties are not yet fully developed (such as green concrete).
- Temporary structures (such as formwork and shoring) are not designed or built with the same care and strength as permanent structures.
- For some systems, the highest loads ever imposed are during initial lifting and erection.

This case is an example of the latter type. Although it is conceivable that the same design error (using separate footings) could have occurred with a permanent structure, the rigidity of permanent components would have prevented the secondary load transfer and buckling that led to the collapse.

Little Changes, Big Problems

In the October 1994 Failures column ("Skin Cancer"), I talked about failed exterior cladding at Park Square Condominiums in Houston, Texas. At that time I focused on cracking and spalling of prefabricated panels due to the absence of reinforcing mesh across joints. Other faulty details contributed to the failure, however. The windowsill configuration was among the worst.

Substituting inferior details for good ones hastened water damage to exterior cladding.

By David H. Nicastro

The drawings to the right depict the as-designed and as-built windowsills at Park Square. There were several significant differences between the two. First of all, the design called for a distinct outward slope in panel returns to the windowsills, but the returns were built level. As a result, water collected on the tops of panels and against the sealant joints under the aluminum windowsills.

Second, the sealant joints under the windowsills were designed to be in the windowsill plane. As installed, however, they were flush with the tops of the panels. Even if the tops of the panels had been sloped, the horizontal sealant joints would not have shed water as well as they would if they had been installed in the vertical plane.

The horizontal configuration also prevented the weep holes from draining the windowsills properly; sealant was found either to have plugged weeps or to be discontinuous under them, preventing water from draining in the one case and allowing water that did drain to reenter in the other.

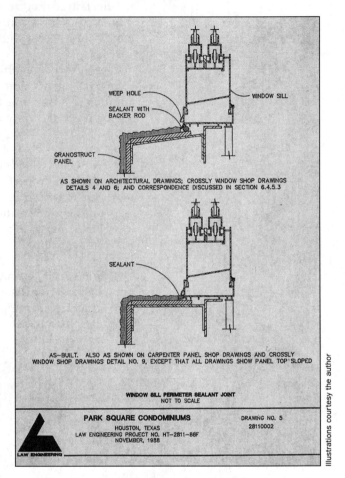

WEEP HOLE
SEALANT WITH BACKER ROD
GRANOSTRUCT PANEL
WINDOW SILL

AS SHOWN ON ARCHITECTURAL DRAWINGS; CROSSLY WINDOW SHOP DRAWINGS DETAILS 4 AND 6; AND CORRESPONDENCE DISCUSSED IN SECTION 6.4.5.3

SEALANT

AS-BUILT. ALSO AS SHOWN ON CARPENTER PANEL SHOP DRAWINGS AND CROSSLY WINDOW SHOP DRAWINGS DETAIL NO. 9, EXCEPT THAT ALL DRAWINGS SHOW PANEL TOP SLOPED

WINDOW SILL PERIMETER SEALANT JOINT
NOT TO SCALE

PARK SQUARE CONDOMINIUMS
HOUSTON, TEXAS
LAW ENGINEERING PROJECT NO. HT-2811-88F
NOVEMBER, 1988
DRAWING NO. 5
28110002
LAW ENGINEERING

Illustrations courtesy the author

Aggravating the problem was the fact that the rough surface of the aggregate panels had made proper tooling of the joint sealant virtually impossible; during tooling, the blade bounced off each stone, leaving small voids. Therefore, adhesion of sealant to the substrate was poor—even where joints appeared to be intact.

An attempt to reseal the joints would have been no more effective and, in any event, would not have resolved the joint configuration problems. As a result, a new exterior insulation and finish system was applied over the old cladding, as described in the previous column. ◆

DAVID H. NICASTRO, P.E., is founder and president of Engineering Diagnostics, Inc., in Houston, Texas. Mr. Nicastro specializes in the investigation and remedy of construction problems and the resolution of related disputes.

Failure Mechanism: Adhesion Failure

The purpose of providing drawings in construction contract documents is to graphically depict the relative position of parts. That goal was resoundingly defeated in this case, as can be seen by comparing the architectural design drawing and the as-built condition of the critical sill detail. In conjunction with the panel cracking, the sealant failures caused pervasive water infiltration and corrosion. It is difficult to determine which defect let in more water—the sealant or the cracks.

The purpose of shop drawings is to verify the detailed design of individual components. That goal was also defeated; some shop drawings showed the change in design was due to the poor arrangement of parts, but the problem was not caught. Worse, the panels were not fabricated in accordance with any of the drawings in one important respect—the slope of the sill, which is necessary to promote drainage.

Rough aggregate-surfaced panels always present a challenge to the waterproofing subcontractor who must seal the joints. First, there must be a smooth return edge (without exposed aggregate) to which the sealant can bond. But even if the internal joint edges are smooth, it is difficult to tool sealant into a joint between substrates with rough surfaces, because the metal tool follows the profile of the surface. Tooling promotes durability of the sealant by forcing adhesion to the joint edges, by filling voids, and by compressing the surface. Inadequate tooling may leave a joint that appears intact, but, in fact, will leak water, as was found here.

On my office wall, I keep a drawing of the critical detail from the Hyatt Regency Hotel pedestrian walkways that collapsed in 1981 in Kansas City, Missouri, killing 114 people. This picture of a small steel connection, which was changed during construction due to a constructability issue, reminds me that there are no small changes; all changes have consequences.

Pop Quiz

The photo depicts a deformed steel bar joist supporting the roof of a school building in Missouri City, Texas. During the summer of 1990, the end weld of several adjacent bar joists suddenly broke during a rainstorm, causing the roof to partially collapse.

In our investigation, we found that a contractor had stored construction materials on the roof in anticipation of replacing the roofing membrane. We also found evidence of ponding. Our observations and calculations indicated, however, that the roof was not overloaded at the time of failure.

Two failed tests mean a summer of remedial work for a Texas school.

By David H. Nicastro

Instead, after examining the bar joists, we found that the welds did not meet industry standards—a particularly alarming discovery, since the same joists were used throughout the two-story building, not just on the roof.

Joist "Populations"

The Steel Joist Institute (SJI) prescribes a method of accepting joists that does not meet visual criteria: if a representative sample passes a full-scale load test, then an engineer can accept the suspect joists without remedy. This approach is based on empirical evidence showing that actual joist strength usually exceeds calculated capacity.

Based on a statistical sample, we developed a model for each "population" of bar joists and designed an in-place load test according to SJI procedures. We gradually filled specially built water tanks 12.2 m (40 ft) long to apply the load. The tanks required internal baffles to keep the load fairly uniform as it was increased; otherwise, all the water would have run to the center of the deflected floor.

We determined that if two or more failures occurred out of any of the joist populations under investigation, then the joists could not be accepted in their inferior condition. Two failures *did* occur from each group, and the joists all received additional welding in place. This remedy required removing most of the interior suspended ceiling.

Despite the magnitude of remediation, joists were strengthened and interior finishes restored before students returned for the fall semester. Even the roof replacement was completed during the summer, as originally planned. ◆

DAVID H. NICASTRO, P.E., president of Engineering Diagnostics, Inc., in Houston, Texas, specializes in the investigation and remedy of construction problems.

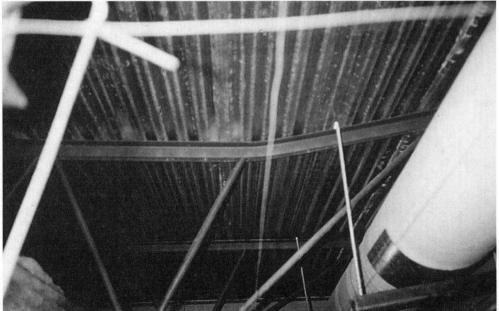

Courtesy the author

Failure Mechanism: Overstress

Steel bar joists are one of the most popular construction components and are found in both low-rise and high-rise buildings. These shop-fabricated trusses often have terrible looking welds, giving certified weld inspectors nightmares. Despite how poor the welds may appear visually, the structural capacity of the joists is usually adequate. In this case, however, they proved to be deficient by full-scale, in-place load testing. The failure was caused by undersized critical end-bar welds, which experience the highest load.

One difficulty in load testing is selecting a statistically representative sample to best characterize the entire population being studied. There are no industry standards for determining an appropriate number of samples, a maximum statistical error margin, a minimum confidence interval, or an assumed distribution (such as "bell curve"). In lieu of a mathematical analysis for each case, I have promoted in the forensic engineering community the following general criteria for statistical extrapolations:

1. Assume that the test results will be normally distributed (i.e., a graph of the data would be shaped like a "bell curve"). Most engineering issues can be reduced to a precise expression in which results would be normally distributed.
2. Use a minimum 90% confidence interval.
3. Use a maximum 20% error margin.
4. If no preliminary data are available, assume the test results will be divided equally among the possible outcomes (a 50/50 chance); this is statistically the worst case and would require the most samples to ensure that the error margin is not exceeded.

One must be careful to explicitly state the question that the testing is intended to answer. A common mistake investigators make is an inappropriate extrapolation to a larger population than was really studied. For example, if all specimens are taken from the west elevation of a building, then statistically only the west elevation can be characterized by the findings, no matter how similar the apparent construction of the other elevations. If the goal is to extrapolate to the entire building, then the samples must be selected from the entire building.

Related to this is the misuse of "random" sampling when "pseudo-random" would be more appropriate. Truly random selection requires using mechanical apparatus (such as dice) to select the specimens, whereas a "random number generator" algorithm can be used for pseudo-random selection. As a practical matter, most investigations are hampered by convenience of access, so the available population to be selected is truncated. In many cases, a more precise term than random or pseudo-random would be "nonbiased"—a broader descriptive term for any sample strategy that is selected without bias as to the specific properties being investigated.

In Chapter 3, I suggested that it is important to pay attention to temporal information. In this case, the timing of the failure raises the question whether it was a coincidence that (1) replacement roofing materials were stored on the roof at the time of the failure, (2) that water ponded on the roof, and (3) that there was a major storm. These factors all imposed loads on the roof and together perhaps created the highest service load yet experienced by the roof. Therefore, they contributed to the failure, but they did not cause it. The forces that acted on the roof at the time of the failure were part of the existing conditions that must be documented and explored, but in this case they did not cause overloading; the bar joists were simply deficient.

The Brick Is Falling

Courtesy the author

Although masonry construction has been around for up to 12,000 years, we occasionally need to be reminded of some basic principles of its proper use. The photo depicts a failure caused by a practice that defies common sense as well as all published works on masonry veneer construction.

The brick faces have spalled off, revealing the toe of a shelf angle at approximately mid-height of the brick course. When I first visited this eight-story university library, ropes had been strung around the building to prevent falling pieces of brick from striking the public.

Masonry courses did not line up with previously erected shelf angles.

By David H. Nicastro

A shelf angle's purpose is, of course, to support the prism of masonry above. Good practice dictates that the bottom masonry course be seated directly on the steel (or rather, on a flashing material, which is also conspicuously absent in this photo). A space should be left between the shelf angle and the masonry course below it. Theoretically, a shelf angle does not take any load until it deflects; so, if it bears on solid masonry below, it passes the loads from above into the masonry below rather than back to the building frame.

In this case, the masonry courses did not line up with the previously erected steel shelf angles. Apparently, to preserve even course lines around the building, the bricks were modified to fit around the steel wherever it occurred. Called "soaps," these cut bricks were typically either L-shaped or thin, flat fronts only. In some areas, there is space below the steel (as in the photo), so the shelf angles did deflect, carrying the masonry above with them. However, the soaps were mortared in solid to the course above, and they were crushed when the weight of the deflecting masonry bore on the thin face cross section.

In addition to the masonry structural failures, the walls leaked like a sieve. Criticisms could be made about the mortar joint profiles (raked joints have significantly lower resistance to water penetration than concave joints) and the numerous cracks and spalls in the masonry. However, the obvious problem related to the mislocated shelf angle was that there was no way for a flashing to drain out of the wall. A flashing should terminate outside the masonry, but that is impossible when it is placed at a row of soaps instead of a horizontal joint.

Sharing the Blame

Although the masonry contractor can be easily criticized for these practices, the general contractor also shares the blame for the lack of coordination between the steel and masonry, and no doubt the architect is responsible for the poor detailing of the shelf angles.

I also criticize the conventional practice, as done here, of structural engineers showing the steel shelf angles on their drawings but indicating the masonry only by a phantom line. This is an intentional abdication of any responsibility for the masonry's behavior, yet engineers are the best hope of preventing this type of failure. The structural drawings should show the steel/masonry coordination. To continue not doing so is to condemn more buildings to this type of failure. ◆

DAVID H. NICASTRO, P.E., is founder and president of Engineering Diagnostics, Inc., in Houston, Texas. Mr. Nicastro specializes in the investigation and remedy of construction problems and the resolution of related disputes.

Failure Mechanisms: Spalling; Overstress

As with the first column (page 8, "The Too-Short Shelf Angle"), we have an example of lack of coordination between steel shelf angles and brick masonry, but this one is even more egregious. Because it is so common to find shelf angles cut too short or extending too long, one could believe that those errors are within the standard of care of engineers. But any reasonable person would expect this shelf angle detail to fail.

This case also illustrates a fundamental concept in forensic engineering. Although the focus of this book is on failure mechanisms and two mechanisms can be identified for the failure that occurred, they are not important in this case. This is simply a case of poor construction practice; precisely describing the mechanisms does not add to the understanding of this failure. The proximate cause here would be the bastardized detail—not any of the identified mechanisms.

Tempered Glass

Tempered glass often is used as safety glazing because of the assumption that it will break into small, harmless pieces (often metaphorically referred to as "rock salt") that will not result in lacerations. In fact, the definition of tempered glass in ANSI Z97.1, American National Standard for Safety Glazing Materials Used in Buildings, states, "When broken at any point, the entire piece immediately breaks into innumerable small granular pieces."

However, the photo shows dangerous shards of broken tempered glass. The glass suffered repeated spontaneous breakage when used in a monumental skylight system with more than 1,300 lites. All of the lites were eventually replaced with laminated glass to protect the public.

Breakage patterns often include dangerous shards.

By David H. Nicastro and Joseph P. Solinski

Although there were allegations that the glass was not fully tempered, it has been our experience that actual breakage patterns of unquestionably fully tempered glass often include hazardous clumps or shards. Therefore, using tempered glass overhead is dangerous and is not allowed by most current building codes. Unfortunately, this application was allowed by the local building code when this failure occurred in 1984.

There are a number of causes of spontaneous breakage of tempered glass, including:

• inclusions (such as nickel-sulfide) that cause rupturing when heated

• differential thermal expansion from unusual shading patterns

• edge defects from manufacturing or handling

• excessive loading

• missile impact (such as wind-borne gravel from roof ballast)

• contact with metal framing due to poor installation practice.

Point-Impact Loading

Tempered glass performs well for most safety glazing applications (such as side lites next to office doors and sliding glass doors). However, it appears that the glazing industry overextended the use of this material based on a misunderstanding of its actual performance.

Although tempered glass is significantly stronger than non-tempered in uniform loading, it does not resist point-impact loading well. Thus, using tempered glass in full-height office windows above the first story can be dangerous.

Other types of safety glazing (such as laminated or wired glass) are recommended for overhead and full-height glazing applications. Also, a chair rail at mid-height can prevent someone from falling through broken full-height glazing.◆

DAVID H. NICASTRO, P.E., is founder and president of Engineering Diagnostics, Inc., in Houston, Texas. Mr. Nicastro specializes in the investigation and remedy of construction problems and the resolution of related disputes. **JOSEPH P. SOLINSKI** is president of Stone & Glazing Consulting in Dallas, Texas. Mr. Solinski consults on design, testing, and investigation of exterior wall systems.

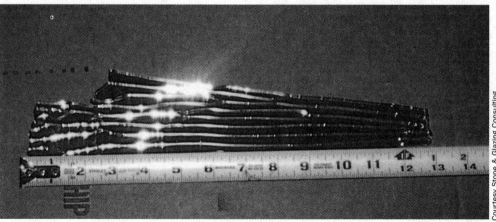

Courtesy Stone & Glazing Consulting

Failure Mechanisms: Spontaneous Breakage; Edge Defects;
Edge Contact; Nickel Sulfide Inclusions; Overstress;
Missile Impact; Thermal Shock

Tempered glass is often called safety glazing, and in most applications its use does significantly increase safety. But tempered glass has been used in some inappropriate applications because of a misunderstanding of its properties. This concept is introduced at the end of the column with a discussion of uniform loading versus point loading. Perhaps the larger misunderstanding is the expectation of tempered glass breaking into "rock salt."

Tempering a lite of glass creates tension and compression layers that are in equilibrium. When ruptured, the lite implodes due to the high (over 10,000 psi) residual stress. The resulting pieces should be not only innumerable and granular but without sharp edges. This is the most important feature of tempered glass as a safety glazing material: it prevents lacerations. But lacerations are not the only hazard from glazing.

Broken tempered glass can clump together rather than totally separating into "small granular pieces." The weight of the clumped glass may cause injury when falling from overhead glazing. Furthermore, there are, in fact, sometimes large, sharp pieces found in the debris after spontaneous breakage. In this case, one of the shards fell from the skylight and stuck into a wooden handrail like a dagger. It was still intact at the time of our investigation, surviving the initial breakage and the impact.

Although the column lists a number of potential failure mechanisms that cause spontaneous breakage, in this case it was concluded that the causes were edge contact and edge defects. The glass was delivered with imperfections along the edges large enough to cause latent breakage. The glazer compounded the problem by allowing the metal skylight frames to touch the edge of the glass, which caused stress concentrations.

The Fish Bowl Roof

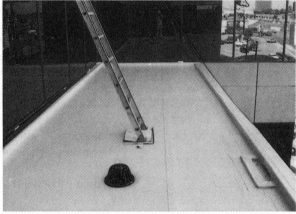

The photos show a roof before and after remedial construction to alleviate severe water infiltration. It is one of three small inset roofs on an office building in San Antonio, Texas.

Each roof originally was constructed with a ballasted built-up roof membrane flush with the top of the window sills; thus, the weep holes in the bottom of the sills were several inches below the membrane. Water in the curtain wall could not drain out onto the roof as intended and leaked to the interior. Because there was no slope to the roof drain, water cascading down six floors of curtain wall accumulated on the roof membrane against the bottom window lites, creating a "fish bowl" appearance from the interior during storms. The water also leaked in through the window gaskets, which were never intended to be submerged.

Adequate drainage is the most important element of waterproofing design.

By David H. Nicastro

Most roofing membrane manufacturers require a 200 mm (8 in.) "freeboard" or curb height to allow drainage from vertical surfaces onto the roof. Here, the freeboard drainage was nonexistent; in fact, the vertical surfaces extended below the roof membrane.

Our remedial design called for removing the membrane and insulation down to the structural concrete deck, which, fortunately, was below the window sills. We then used sloped isocyanurate insulation to create minimal but positive drainage to the roof drain, which had to be lowered from the original membrane height down to the deck level. A single-ply membrane was fully adhered to the insulation. The design allowed for minimum total thickness at the perimeter curbs, so that maximum freeboard could be established.

Metal flashings were adhered with structural silicone sealant directly to the window lites and mullions above the sills to provide a waterproof transition from the vertical surfaces to the new roof membrane. The windows above were changed from a drainage-type to a barrier-type system—since water was no longer allowed to drain out of the bottom weep holes, it had to be prevented from entering the system in the first place. All window lites directly above the roofs were wet sealed by applying a heel bead of silicone sealant over the window gaskets and "bridge seals" over the metal-to-metal fit-up joints in the aluminum framing.

The completed remedies have been functioning since 1986 without leakage, once again demonstrating that adequate drainage is the most important element in waterproofing design. ◆

DAVID H. NICASTRO, P.E., is founder and president of Engineering Diagnostics, Inc., in Houston, Texas. Mr. Nicastro specializes in the investigation and remedy of construction problems and the resolution of related disputes.

Failure Mechanism: Not Applicable

This column states that adequate drainage is the most important element in waterproofing design. For a discussion of the necessary elements, see the commentary on the March 1994 column "The Deck Drain That Doesn't." Another general observation that can be abstracted from this column is the ubiquity of water infiltration problems associated with plane transitions from horizontal to vertical.

There are several practical difficulties at horizontal/vertical transitions that may account for the pervasiveness of defects:

- The horizontal and vertical components are usually constructed at different times and may be waterproofed separately and by different contractors. Therefore, the two systems may not be integrated. In this case, the glazer was responsible for the vertical waterproofing and the roofing subcontractor for the horizontal.
- There are stress concentrations at changes in plane, which may cause movement or overstressing of materials.
- Water naturally travels along horizontal surfaces until interrupted by a change in plane; thus, there is an accumulation of water at these critical details.

A "Simple" Wall Failure

Sometimes a forensic engineer's job is easy: the cause of a construction failure seems so obvious during an investigation that it is hard to believe that the defect was not noticed and corrected. Such was the case with the "simple" wall failure shown in the diagram.

Concrete masonry unit (CMU) infill walls are often used in industrial facilities to separate bays for fire safety or to separate functions such as office space from warehouse. Often these walls are added to an existing facility and are not designed by a structural engineer, because they are not intended to resist or transfer building loads. Regardless of the intent, CMU walls can carry significant vertical forces (gravity loads), and most building codes also require them to be designed to carry nominal lateral loads.

In this case, a wall was installed between floors of an existing paper mill building without considering the effects of vertical load transfer from the floor above through the new wall onto the supporting structure below. The masonry was installed tight against the bottom of the girder above. When the tanks above were filled, the live loads were passed through the wall into the girder below and caused the riveted seat connection to fail. At the time of failure, it is estimated that the seat connection was loaded to two and a half times its allowable design load.

Catastrophe was avoided because displacement of the girder end redistributed the load to the adjacent framing. If the transferred load had not been redistributed after the initial failure, a progressive failure would likely have ensued, collapsing a significant portion of the structure.

To remedy the situation, the failed seat connection was repaired, and the top of the masonry wall was removed to eliminate transfer of live loads on the floor below. The space was filled with compressible insulation, and brackets were installed to laterally brace the top of the wall.

Monday Morning Quarterbacks

When the cause of a failure seems most obvious, forensic engineers must avoid the trap of believing that we would not make such mistakes if we were the designers. The number of failures resulting from design errors demonstrates that they are sometimes easier to diagnose than prevent. ◆

> ### Failures from design errors are sometimes easier to diagnose than to prevent.
>
> ### By David H. Nicastro and Patrick R. McCormick

DAVID H. NICASTRO, P.E., is founder and president of Engineering Diagnostics, Inc., in Houston, Texas. **PATRICK R. McCORMICK, P.E.,** is a senior engineer with Brander Construction Technology, Inc., in Green Bay, Wisconsin, and an active member of ASCE's Technical Council on Forensics and its Committee on the Dissemination of Failure Information.

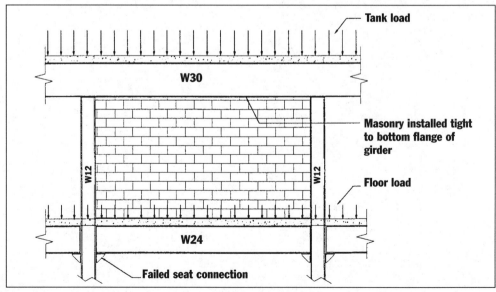

Failure Mechanism: Overstress

A certain degree of humility is necessary in forensic engineering, lest we begin to believe that we are immune from causing failures simply because we have expertise in analyzing them. I have previously alluded to the drawing, which I keep on my office wall, of the critical detail from the Hyatt Regency Hotel pedestrian walkways that collapsed in 1981 in Kansas City, Missouri, killing 114 people (see commentary to "Little Changes, Big Problems", page 33). This picture reminds me "there but for the grace of God go I," because there are so many opportunities to err in structural design.

In this case of a "simple" wall failure and in the Hyatt Regency case, it is easy to understand why the failures occurred, which may lead us to believe that we would not have made the mistakes that caused them. I disagree; both mistakes were easy to make, even though they were errors in engineering judgment. In both cases, I believe that better design policies, not skill, would have prevented the failures. It is not reasonable to expect structural designers to scrutinize every detail with insight, but it is reasonable to require them to use redundant load paths, ductile materials, and standardized details wherever possible. In other words, if the designer of this infill wall had a policy of always leaving a gap above masonry as a standard practice (a good idea for a number of reasons), then he/she would not even have had to consider the unintended load path that developed.

Tower Distress

What happens when you stretch steel framing tightly between two rigid towers 60 m (200 ft) apart? In the five-level parking garage pictured, the towers suffered severe distress.

The concrete slabs are supported by steel framing that engages stair towers built of concrete masonry and brick veneer. Severe cracking and displacement of the masonry occurs where the steel frames fit into the stair towers. Similarly designed structures throughout the country often suffer the same fate.

Cracks and displacements occur where the frame meets the masonry.

By David H. Nicastro and Kimball J. Beasley

Thermodynamics

The vast majority of performance failures caused by thermal expansion and contraction result from major misconceptions in the industry regarding thermodynamic influences on structures. Some often misunderstood factors include:

• An exposed building component responds to its own internal temperatures, not to the ambient temperature. Changes in internal temperature cause a combination of stress and strain.

• There is a time lag between an air temperature change and a structure's response, based on the "thermal mass" of the system, material properties, and insulation.

• The actual temperature range that the structure experiences significantly exceeds the ambient temperature range. An exposed component can become significantly hotter than the hottest ambient air temperature due to solar gain. An exposed component can become colder than the coldest ambient air temperature (dry bulb) due to radiative cooling, and because porous materials can experience evaporative cooling.

Expansion Joints

In the investigation of the garage, calculations indicated that substantial lateral movement of the exposed portion of the steel frame were caused by seasonal and daily temperature changes. If analyzed as fully unrestrained (so that all thermal effects would result in movement rather than stress), over 25 mm (1 in.) of movement would occur between the stair towers seasonally. The relatively rigid masonry stair towers were incapable of accommodating this much frame move-

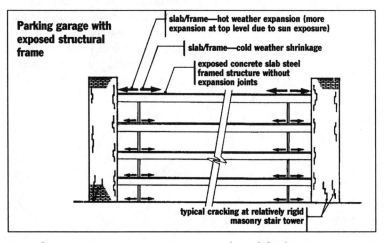

Parking garage with exposed structural frame

slab/frame—hot weather expansion (more expansion at top level due to sun exposure)

slab/frame—cold weather shrinkage

exposed concrete slab steel framed structure without expansion joints

typical cracking at relatively rigid masonry stair tower

ment, causing cracks and displacements to occur where the frame meets the masonry.

The stair tower masonry must be partially or completely rebuilt with expansion joints. However, the remedial design must consider that the new expansion joints may alter masonry load paths and lateral stability. Additional masonry supports and anchors may be required.

In structures where most or part of the structural frame remains exposed to exterior temperature variations the need for judiciously located expansion joints is more acute than in structures where the frame is sheltered. The frame must accommodate all anticipated movements and load-related deflections without damage to stiffer elements, such as masonry walls. ◆

DAVID H. NICASTRO, P.E., is president of Engineering Diagnostics, Inc., in Houston. He specializes in the investigation and remedy of construction problems and the resolution of related disputes. **KIMBALL J. BEASLEY, P.E.,** is a senior consultant with Wiss, Janney, Elstner Associates in Princeton, New Jersey. He evaluates serviceability failures of building components.

Failure Mechanisms: Cracking; Thermal Expansion; Overstress

Notice the similarity between this case, involving a structural concrete frame between masonry towers, and the case described in "Of Stress and Stucco," page 12, involving metal handrails between stucco walls. The resemblance is striking, even though one relates to the macroscopic behavior of an entire building, whereas the other is just the localized behavior of a single member.

Recognizing this scalability of failure mechanisms can be a useful technique in forensic engineering. I once observed an unusual cracking pattern in an enormous concrete water tank—large, nearly vertical cracks at regular intervals around the tank. Standing inside the tank, I noticed how similar the appearance was to an old barrel made up of wooden slats with metal hoops. This was the clue that made me realize (and later prove) that the tank had inadequate hoop reinforcement; it was separating into "slats," which were falling outward.

Stone Stains

Fortunately, most construction failures are not catastrophic. The majority of disputes over building materials and systems that fail prematurely involve rather mundane nuisances, such as staining of exterior walls. But non-life-threatening problems are not necessarily inexpensive or easily solved; I have clients who have spent millions of dollars to alleviate aesthetic problems on high-rise curtain walls.

One cause of aesthetic failures is granite staining (see photo). Water absorption caused the stains at the base of this skyscraper. Stone cladding can become water stained if it is continually wetted from a cut edge. This occurs adjacent to joints sealed with a water-absorptive product—in this case, a polyurethane-based sealant with an open cell backer rod. The solution: Replace the sealant with a nonabsorptive (and nonstaining) product. The stone will then dry on its own.

Fluid migration from the sealant to the substrate can cause staining.

By David H. Nicastro

Loose Juice

Stains also occur next to elastomeric sealant joints. Low viscosity fluid that migrates from the sealant into the substrate is a phenomenon sometimes referred to in the sealant industry as "loose juice." Fluid migration is most often associated with low-quality silicone sealants, but it can occur with other types of sealants adjacent to a porous substrate.

After a sealant absorbs into the pores of a cladding material, aesthetic problems can occur. The sealant can do any of the following:

• migrate to the surface and attract and hold airborne dirt, which is difficult to wash off because of the chemical bonding to the surface

• waterproof a portion of the cladding in a zone near the joints, repelling normal wetting during rainfall and leading to a checkerboard appearance

• darken the substrate, giving a wet appearance near the joint.

The zone of fluid migration is usually less than 20 mm.

Other Causes of Staining

Sealant is not used only in joints; stone anchors also are often set with sealant, which allows fluid migration to the stone's face near each anchor, even in the middle of a panel. Perhaps the most disturbing fact about these common aesthetic problems is that standardized test methods commonly specified do not always accurately predict the latent development of staining.[1]

Fluid that migrates out of a sealant does not always absorb into the pores of a substrate; it can also wash down onto the surface. In such cases, the affected zone can be much larger than 20 mm, but would otherwise mimic the effects of fluid absorption. Recently, experimental remedies for these phenomena have been introduced. Generally, a poultice that digests the offending fluid is applied to the affected surface.

Natural stone near elevator buttons, door handles, or other human contact areas tends to absorb body oils. This phenomenon is easy to diagnose by its pattern, and the remedies are similar to those for fluid migration from sealants. ◆

Note

1. V.K. O'Neil, et al., "Effects of Weatherproofing Sealants on Building Aesthetics," Science and Technology of Building Seals, Sealants, Glazing and Waterproofing, ASTM STP 1243, vol. 4, David H. Nicastro, ed. (Philadelphia: ASTM, 1995).

DAVID H. NICASTRO, P.E., is founder and president of Engineering Diagnostics, Inc., in Houston, Texas. Mr. Nicastro specializes in the investigation and remedy of construction problems and the resolution of related disputes.

Failure Mechanisms: Fluid Migration; Water Absorption; Rundown; Freeze-Thaw

One of the most prevalent aesthetic problems we see with stone-clad buildings is water absorption into the stones from the sealant joints. This is not only an aesthetic problem, because the absorbed water can cause freeze-thaw damage.

The cut stone edges are usually much more absorptive than the face. Both flame finishing and polishing introduce heat into the face, which fuses the grains together, making a fairly impervious membrane. The unfinished edges are naturally more vulnerable and made more so by the cutting operations. Placing absorptive materials (open-cell backer rod and some sealants) against the stone edges creates a reservoir to continually feed water into the stones.

The column lists other aesthetic problems associated with stone joints, but they can easily be differentiated from water absorption, which varies in appearance as the water absorbs and evaporates; fluid migration and rundown only get worse.

Sealant Reversion

ASTM defines a sealant in building construction as "a material which has the adhesive and cohesive properties to form a seal." Unfortunately, this does not describe the condition of the elastomeric joint sealant in the photo. The sealant has lost its original rheological properties—it no longer is a cured rubber, but instead has become soft and non-elastic with the consistency of chewing gum.

This type of sealant failure is called *reversion*. Although there is no written history of the development of this phenomenon, I first heard about reversion in the mid 1980s as a contractor's rebuttal to allegations of improper workmanship causing sealant failure. Two common claims against contractors that could cause similar manifestations are:

• using out-of-shelf-life products; however, this could be easily prevented by checking the label.

• improper mixing of multi-part products; however, components are rarely the same color, so improper mixing tends to leave striations or marbling in the joint.

Recognizing that there was a new phenomenon in numerous projects and that they were not at fault, contractors described the condition not as a failure to cure, but as reversion—the sealant had initially cured and then *reverted* to an uncured state. (Both of the mistakes listed above would prevent the sealant from curing in the first place.)

The word reversion is also used in polymer chemistry to describe a similar phenomenon; however, it is not necessarily precise when applied to construction sealants. Many practitioners believe that the failure mechanism represents a new state, not a reversal of the chemistry to an earlier state.

Reversion has been a controversial subject in the sealant industry because it is not well understood. Complaints of reversion have been primarily associated with urethane-based sealants and have become so commonplace that some products now advertise that they will not revert.

Scientists studying the problem have not yet reached a consensus on the causes: in fact, laboratory studies and field data have been contradictory. It is generally recognized, however, that the failure involves high temperatures, ultraviolet (UV) exposure, and water or vapor working alone or in combination.

> The sealant initially cured and then reverted to an uncured state.
>
> By David H. Nicastro

Courtesy the author

The majority opinion appears to be that high temperature alone can cause reversion in the field, but a dissenting minority believes water absorption is essential, and one study concluded that UV was the dominant factor.

One theory is that reformulations in the early 1980s reduced the weathering resistance of some products, giving rise to a rash of failures. New cases are still being discovered every day.

Diagnosing Reversion

Reverted sealant can be easily diagnosed by removing a small sample—if it is known that the product's shelf-life has not expired. If no striations are visible and the sample can be rolled into a ball that holds its new shape, then it has lost its rheological properties and can be referred to as reverted.

Interestingly, the photo also dispels a common myth about sealants: that painting over them is always detrimental. Obviously, the upper painted material is performing better than the unpainted material below. Recent field data from a variety of projects confirms that painting can improve the durability of many high-performance sealants, probably by reducing the UV exposure.[1] ◆

Note

1. P.D. Gorman, "Weathering of Various Sealants in the Field—A Comparison," *Science and Technology of Building Seals, Sealants, Glazing and Waterproofing*, vol. 4, ASTM STP 1243, David H. Nicastro, ed. (Philadelphia: ASTM, 1995).

DAVID H. NICASTRO, P.E., is founder and president of Engineering Diagnostics, Inc., in Houston, Texas. Mr. Nicastro specializes in the investigation and remedy of construction problems and the resolution of related disputes.

Failure Mechanism: Reversion

Although reversion falls within the scientific discipline of rheology (the study of flow), defining it as a loss of rheological properties was not the best choice of terminology. A more descriptive definition is the loss of elastomeric properties— the inability to rebound like rubber after being deformed.

Reversion has a colorful, albeit brief, history in the sealant industry. First, the phenomenon was denied to exist; second, it was blamed on a variety of causal factors in a very unscientific manner; third, it was blamed on multiple factors occurring simultaneously; and finally, it is now acknowledged by manufacturers who advertise that their products do not revert.

So what does cause reversion? Probably several things, depending on the specific organic sealant formulation. Even the color of the sealant is a factor, because of the energy absorbing properties of different pigments. Various reports of laboratory testing studies have contradicted one another. Field data, however, indicate that joints with high heat accumulation are vulnerable and exacerbated by water accumulation.

The column dispels the myth that painting is detrimental to sealant performance. Paint applied over a sealant can induce cohesive failure by propagating a crack from the brittle coating into the soft sealant, and paint also can introduce alcohol or other solvents into fresh sealant, inducing adhesive failure or inhibiting cure. Nevertheless, much field data indicate that these potential problems may be more than offset by the benefit of protecting an organic sealant from ultraviolet light, the primary factor in weathering.

Glazing System Leaks

Most glazing in commercial construction uses perimeter gaskets around each glass lite. Lock-strip gasket glazing, which has no exterior metal framing, and conventional dry compression glazing in aluminum frames both rely on compression of the exterior gaskets to maintain a weatherseal and to structurally retain the glass. Glass manufacturers recommend a gasket lip pressure of 720 N/m to 1760 N/m (4 pounds per linear inch [pli] to 10 pli), but what happens if the lip pressure falls lower?

Most gaskets are made of elastomeric materials that degrade over time due to ultraviolet radiation and ozone exposure. External gaskets are subject to extreme exposure, not only from direct sunlight but from reflected radiation as well.

As the volatile components are driven off, gaskets may shrink in volume or become brittle with age—both of which limit their ability to seal. In a few cases, the gaskets may have to be replaced to prevent glass loss in high winds. A much more prevalent result is water infiltration.

Weathered gaskets may not provide adequate seal against the glass.

By David H. Nicastro

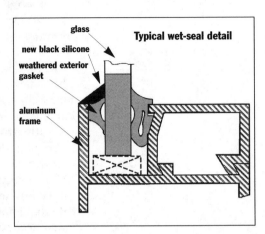

Typical wet-seal detail

glass
new black silicone
weathered exterior gasket
aluminum frame

Water Leakage

Even though conventional wisdom has held that a seal pressure of 1080 N/m (6 pli) is necessary to prevent leakage, empirical field studies have shown that gasket lip pressure as low as 440 N/m (2.5 pli) can prevent water infiltration. However, the lower the lip pressure, the more vulnerable the system is to small discontinuities along the gasket/glass interface. It is also particularly difficult to measure the in situ lip pressure; some devices have been made for this purpose, but they disturb existing conditions to make measurements.

Except in barrier systems, which rely solely on the outermost components to prevent leakage, most designs permit some water to percolate through the gaskets. The water is then extricated through weep holes. However, the internal seals in these glazing systems are often defective and cannot be remedied without deglazing, which is costly, time consuming, and disruptive to tenants. Preventing water entry through the weathered gaskets is a desirable remedy.

Wet Sealing

Wet-sealing (applying a heel bead of sealant over the exterior gaskets) is one of the most commonly implemented remedies. Yet it is often incorrectly performed. In countless cases, specifying the removal of a previously failed heel bead before wet-sealing has been required, with the added burden of convincing the owner to spend money on the same repair again.

When properly done, as shown in the sketch, wet-sealing can last for the life of a building and not detract from its appearance. As a collateral benefit, wet-sealing can improve the structural performance of the glazing. The critical components of a successful application include:

• Use only high-quality silicone sealant, and test for compatibility with the gasket and other contact materials.

• Use black sealant installed by skilled applicators so that the heel bead will appear like a gasket from inside or out.

• Solvent clean and then prime all surfaces, including glass, metal, and old gaskets.

• Maintain a minimum 6 mm (¼ in.) bite, or adhesion contact area, to each surface and a minimum 6 mm cross-sectional thickness.

• Seal fit-up joints or splices in the frames as well; they typically admit a larger volume of water into the same cavities that the gaskets do.

In controlled tests, water enters more readily at the corners of windows than along the straight sides. Therefore, an appropriate remedy may be to simply wet-seal the gaskets in the corners, although this will not save much money and the remedy will probably be more visible than a continuous heel bead. ◆

DAVID H. NICASTRO, P.E., is founder and president of Engineering Diagnostics, Inc., in Houston, Texas. Mr. Nicastro specializes in the investigation and remedy of construction problems and the resolution of related disputes.

Failure Mechanisms: Inadequate Lip Pressure; Out-Gassing;
Shrinkage; Embrittlement; Weathering

Due to space limitations, several additional considerations for wet-sealing
were omitted from this column.

- Generally, the existing gaskets should not be cut back, which might expose
  incompatible (nonvolatilized) inner rubber to the fresh sealant.
- After wet-sealing, glass replacement can only be performed from the exterior. Most owners, however, do not believe that this imposes an undue
  hardship, as many buildings were designed to be reglazed from the exterior in the first place. Spandrel glass, if not vision glass, is almost always
  glazed from the exterior.
- The wet-seal slightly increases the edge shadowing on the glass, which can
  theoretically cause breakage due to thermal shock in a marginal system.
  Generally, maximum thermal stresses occur when <25% of the glass area
  and >25% of the perimeter length is shaded.
- Finally, a dilemma arises in most wet-sealing projects: whether to seal the
  weep holes in the window frames. My experience indicates that it is necessary to seal them, which is consistent with converting the glazing into a
  barrier system. But I only reluctantly recommend this procedure, in light of
  the numerous projects that have had their weep holes plugged as a sole
  remedy, causing worse problems (see page 26, "Wrong-Headed Remediation").

Flashing Failure

At first glance, this photo of a brick masonry wall excavation might appear to reveal satisfactory construction details. So why does this building leak like a sieve?

Courtesy the author

Leak-Free Design and Construction

The wall is an example of a cavity wall: the exterior brick masonry is a veneer over a back-up wall (composed of concrete masonry or steel studs at various places in this building). Like the "rain-screen principle" in modern curtain walls, a cavity wall does not require the outermost components to be absolutely weathertight because there is a back-up system.

> ## Water can run under flashing that does not extend past the wall.
> ## By David H. Nicastro

The Brick Institute of America (BIA) recommends a minimum 50 mm (2 in.) wide cavity to prevent mortar droppings from clogging the cavity, creating mortar bridges between the veneer and back-up wall and blocking the weep holes. Although the cavity in the photo is narrower and substantial mortar droppings were found, the droppings cannot be the cause of the pervasive leaking observed.

Masons almost always completely fill the bed joints but often only partially fill the head joints with mortar. This allows more water to penetrate the veneer, but that is only a difference of degrees—some water is expected to penetrate and drain down the inside face of the veneer.

In the photo, the joints are reasonably well filled. They are also tooled concave, which is the best mortar joint profile for shedding water off the face of the wall, and there is not an excessive amount (i.e., more than 15 percent) of joint separations (shrinkage cracks).

Weep holes are on 610 mm (24 in.) centers, as BIA recommends. They are fully open head joints, which drain better than weep tubes, and are baffled to prevent insects from nesting inside them.

Faulty Design

In the photo, the flashing, which directs water from the cavity to the exterior through the weep holes, is still elastic. Unreinforced PVC flashings are notorious for developing holes, embrittling, and shrinking.[1] The flashing in the photo is in good condition because of the product type, and the seams are well lapped.

The top edge of the flashing does not terminate in a reglet, as would be preferred, but it is still well adhered to the back-up wall with mastic.

However, the flashing has a fundamental design error.

Notice that the front edge of the flashing terminates 13 mm (½ in.) inside the wall, which was required by the specifications and drawings but is discouraged by BIA. It can allow water to curl around the toe of the flashing and run back inside under it.

On this building, we performed systematic water testing and demonstrated conclusively that the source of the leakage is simply because the flashing does not extend past the face of the wall.

Another unusual detail is the use of a grout dam: the flashing is raised one course and rests on the mortar-filled collar joint. Use of the grout dam leaves one course of masonry around each entire floor with no cavity, flashing, or weep holes. No doubt this detail exacerbates the leakage. ◆

Note

1. Clayford T. Grimm, "The Hidden Flashing Fiasco," *The Construction Specifier*, June 1994.

DAVID H. NICASTRO, P.E., is founder and president of Engineering Diagnostics, Inc., in Houston, Texas. Mr. Nicastro specializes in the investigation and remedy of construction problems and the resolution of related disputes.

Failure Mechanisms: Embrittlement; Shrinkage

Prior to this case, I had an academic understanding of the importance of having flashings extend beyond the face of the wall; this recommendation is widely published and is logical. But is it really necessary?

Our systematic water testing and dissection of this wall confirmed beyond a doubt that the voluminous water infiltration was caused primarily by the short flashing leg. Mortar droppings and the grout dam detail also contributed to leakage in the building, but the water flow around the toe of the flashing was quite dramatic. Many architects prefer not to have the flashing visible on the exterior, so they would have been pleased if this turned out not to be critical.

To correct this defect, the flashing must be replaced, which entails temporarily removing several courses of masonry at each shelf angle. This is a radical, invasive remedial procedure, but the leakage is too severe to be arrested by miscellaneous masonry repairs. The design challenge is to develop a cost-effective method of supporting the masonry while courses are removed.

Masonry/Plank Construction

Buildings constructed of prestressed, precast, hollow core, concrete floor planks with load-bearing masonry walls have gained popularity over the last 20 years due to their economy and speed of construction. However, numerous failures have occurred because of a few recurring bad details.

The photo shows a masonry/plank building constructed in 1985 with separating floor planks caused by structural movement, which severely cracked the exterior masonry veneer.

The exterior walls of masonry/plank buildings

Repairing existing masonry/plank buildings is possible.

By David H. Nicastro and Kimball J. Beasley

typically are brick veneer backed with concrete masonry and are designed for composite behavior with grouted (or "slushed") collar joints. The floor planks usually bear on the concrete masonry backup wythe. The following are three common problems with this type of construction.

Incompletely filled collar joints. While the building's structural design relies on masonry composite action, often the collar joints (between the wythes) are not properly grouted. Because the collar joints are hidden after construction, such a deficiency may go unknown until a failure develops. Poorly filled collar joints can lead to substantially diminished masonry load-bearing capacity (especially when the floor planks bear only on the concrete masonry) and reduced lateral wall stability.

Also, the resulting void spaces within the wall can collect water. This can eventually lead to the interior; evaporate through the wall surface, causing efflorescence; freeze, damaging the masonry from expansion forces; or stay concealed within the wall, supporting organic growth and deteriorating the construction materials.

Inadequate diaphragm action. With masonry/plank structures, floor and wall diaphragm action must accommodate lateral loads (e.g., wind or earthquake). To develop floor diaphragm action,

Courtesy the author

continuity between planks with connections or a reinforced concrete topping is necessary. In many cases, only a thin "flash patch" is applied over the planks (as shown in the photo), and a mortar shear key connection is provided between the planks to restrict differential deflections. But a key or flash patch will not assist in developing diaphragm action.

Lack of horizontal expansion joints in masonry walls. The load-bearing nature of masonry/plank buildings makes it difficult to include horizontal expansion joints in exterior masonry walls. Differential movements can occur between the brick veneer and concrete masonry backup due to thermal strain, shrinkage, and creep of the concrete masonry and irreversible moisture growth of fired clay bricks. Plank displacement, masonry cracking, and instability may occur where the floor planks engage the expanding brick cladding (as in this case).

Preventive Design

Masonry/plank buildings should be designed so that the floor planks are adequately connected to each other and to the supporting walls, the masonry is fully composite, and differential expansion of the brick veneer is minimized and accommodated. A diligent quality assurance program during construction is essential.

Repairing existing distressed masonry/plank buildings is possible. Installing connections between planks or a concrete floor overlay may be required to enhance diaphragm action. Stiffening elements can be added where stability is questionable.

Differential masonry expansion may be reduced by structurally isolating the floor planks from the veneer, but a structural analysis should be performed because of the altered load paths.

In this case, connections were added between the planks, and new expansion joints were cut in the exterior brick veneer. ◆

DAVID H. NICASTRO, P.E., is founder and president of Engineering Diagnostics, Inc., in Houston, Texas. **KIMBALL J. BEASLEY, P.E.,** is a senior consultant with Wiss, Janney, Elstner Associates in Princeton, New Jersey.

Failure Mechanisms: Differential Movement; Shrinkage; Creep;
Thermal Expansion; Moisture Expansion; Cracking; Organic
Growth; Deterioration; Ice Crystal Formation; Efflorescence

I believe that the primary reason that failures are so prevalent is the false economy of cheap construction. Masonry/plank construction is popular because it is inexpensive, but there are hidden costs. If the cost to correct the almost inevitable failures were included in the original construction cost, then this problematic system may not really be less expensive than more conventional structures. It is instructive to note how many failure mechanisms are listed in this column.

- Differential movement
- Shrinkage
- Creep
- Thermal expansion
- Moisture expansion
- Cracking
- Organic growth
- Deterioration
- Ice crystal formation
- Efflorescence

A wise professor taught me to pay attention to when I was trying to solve too many problems with one system; changing fundamental design concepts may be more effective than solving each problem.

Frame Shortening

The 24-story tower was constructed in 1983 and, shortly after being occupied, suffered severe distress in the exterior brick masonry veneer. The timing and accumulation of distress toward the bottom of the tower as shown in the photo are both characteristic manifestations of the failure mechanism: frame shortening.

The tower frame was constructed of conventionally reinforced cast-in-place concrete columns and post-tensioned floors. Although it is well known that concrete shrinks over time, it is not often recognized that the resultant shortening of a tall building frame can wreak havoc on the exterior curtain wall if not accommodated in the design.

Courtesy the author

The majority of the strain occurs early in the life of the structure.

By David H. Nicastro

Causes of Distress

In the photo, notice that the courses are out of alignment across the vertical expansion joint (at card); this is most evident in the top mortar joint. Also, the shelf angle no longer supports the masonry to the right of the expansion joint, except for the one brick that has fallen down to rest on it.

When constructed, the masonry was fully seated on the shelf angles, and the courses were in alignment. Tremendous in-plane forces have distorted the steel angles and translated the prism of masonry downward to the left of the expansion joint, leaving the masonry to the right behind. Other distress included spalled and cracked brick units, stair-step cracks through the masonry, and out-of-plane displacement and rotation of entire masonry panels.

Column shortening is caused not only by shrinkage, but creep and elastic compression as well. Applying the dead load from the construction of each additional floor causes the columns below to compress and then to creep over time, in addition to continuing to shrink due to hydration of the port-

land cement. While creep and shrinkage continue indefinitely at an ever-decreasing rate, the majority of the strain occurs early in the life of the structure.

Although the top floor is displaced the greatest vertical distance from its intended position (each floor below contributes some), the greatest individual column shortening occurs at the lowest floor. Because the shelf angles are anchored to the building frame, the lower floors squeeze the masonry, accounting for the characteristic accumulation of distress at these floors.

If horizontal expansion joints are incorporated below each shelf angle, they can accommodate frame shortening as well as other in-plane masonry movements. Unfortunately, no horizontal joints were specified in this project.

The calculations to quantify the amount of frame shortening (either as a prediction for design or as an analytical estimate in a forensic study) are extremely complex, involving many embedded equations. Frame shortening is a function of the time of placement, curing, volume/surface ratio, and amount of reinforcing of each column as well as the time of application of each additional load (floors constructed above). We used a multidimensional array to input all of the detailed data and perform these calculations for this project, which confirmed that the strains were excessive. ◆

DAVID H. NICASTRO, P.E., founder and president of Engineering Diagnostics, Inc., in Houston, Texas, specializes in the investigation and remedy of construction problems and the resolution of related disputes.

Failure Mechanisms: Frame Shortening; Moisture Expansion; Shrinkage; Creep; Cracking; Spalling; Differential Movement

All building frames compress somewhat as they are erected, simply due to elastic shortening. Concrete frames also experience shrinkage and creep, which add to the elastic shortening, and in tall buildings can affect floor levelness and cause distress in exterior cladding, elevator shafts, and other continuous systems.

While the concrete frame shrinks, the masonry veneer grows, due to moisture expansion. Coupling these two systems magnifies the differential movement, resulting in extreme compression of the masonry veneer and consequential distress.

As a worst-case scenario, this building was constructed on a fast-track schedule; the frame was erected at a pace of about one floor per week, with the masonry veneer following a few floors below. Therefore, the two systems were mated together before most of the frame shortening or the masonry moisture expansion had occurred.

On a companion building that suffered a similar fate, new horizontal expansion joints were cut below the shelf angles to relieve the residual stress. The masons reported that the new expansion joints would repeatedly close up as soon as they were cut, because of the rebounding veneer.

Crumbling Concrete

My firm recently removed large pieces of concrete from the precast concrete panels of a 20-story building (see photo). Although none fell out on their own, the pieces were loose and were removed using only plastic mallets. What is most interesting about this case is how rapidly the deterioration occurred.

The Second Time Around

The building was constructed in 1960. When the facade was evaluated about seven years ago, the findings indicated that the precast concrete units were beginning to suffer from environmental exposure. A few spalls were

Courtesy the author

removed, and the cracks were routed and patched.

During a recent follow-up, we intended to use the mallets to create a sonic response for a routine auditory survey of the concrete condition. However, we found the concrete to be so deteriorated that we were able to remove hundreds of pieces. Most of the concrete spalls were found near the edges of the panels. Some were very deep and required the removal of the entire concrete sill.

Unfortunately, a key recommendation from the previous study was not implemented. While the existing distress was repaired, the panels were not waterproofed with a penetrating sealer, which would have reduced or eliminated further deterioration.

The observed distress is predominantly water related. Several contributing failure mechanisms are likely:

- freeze-thaw (repeated formation of ice crystals causes microcracking)
- corrosion of embedded steel reinforcement (the corrosion product is larger than the base steel, causing internal pressure)
- algae growth in cracks (which propagates them)
- debris carried into cracks by water (which wedges the cracks open wider)
- acid rain (pollutants carried in by rainwater become more concentrated as the water repeatedly evaporates).

Originally, the panels were poorly fabricated with voids, honeycombing, and weak structural details at the narrow fins and embedded weep tubes. We found a plaster-type material and wood in some spalls, apparently used to fill voids and grouted over to achieve the intended shape of the panels.

The concrete's porosity also allows water to migrate easily through the panels and leach an alkali solution that stains the windows. The glass lites can be mechanically buffed with a cleanser to improve the optical quality of the vision lites and prevent the glass from being etched.

The Solution

Not all of the loose concrete could be removed during the study, so immediate comprehensive remedial work is scheduled. This will involve using power hammers to remove incipient spalls, reconfiguring panels with cementitious material, and waterproofing panels with a penetrating sealer. ◆

DAVID H. NICASTRO, P.E., founder and president of Engineering Diagnostics, Inc., in Houston, Texas, specializes in the investigation and remedy of construction problems and the resolution of related disputes.

The panels were not waterproofed to reduce or stop further deterioration.

By David H. Nicastro

Failure Mechanisms: Spalling; Cracking; Corrosion; Freeze-Thaw; Algae; Honeycombing

I stated in this column that the deterioration occurred very rapidly. A more complete description would have been that the deterioration was slowly accumulating but became very apparent suddenly. This is consistent with our observations of the behavior of unprotected precast concrete from other buildings, in both northern and southern climates: microcracking, freeze-thaw damage, and corrosion occur continuously, but the degree of distress accelerates as the different mechanisms support each other.

In this case, the time between my two studies was 7 yr, but I have seen even worse distress develop (or rather, become obvious) in less than 2 yr. As the deterioration accelerates, the window of opportunity to arrest the distress becomes shorter. Finally, no remedy but comprehensive replacement would be effective. Thus, a building owner deferring maintenance of apparently minor serviceability problems can be caught by surprise, suddenly being faced with a major structural renovation.

Falling Snap Covers

The drawing at right shows a typical cross section of an aluminum extrusion from a window sill in a curtain wall with an exterior snap cover. As the name implies, exterior snap covers should snap into place on the main sill extrusion. But in the case of a high-rise building, the covers fell off—not just a few but hundreds.

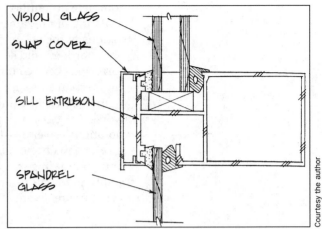

Courtesy the author

Structural Failure

All components of a curtain wall must perform three important functions:

• Resist weather. Because the exterior walls and windows separate outside from inside, they serve as the primary defense against air and water infiltration.

• Maintain the aesthetics of the building. A curtain wall defines the visual statement a building makes; every component contributes to this image.

• Resist and transfer loads. Although a curtain wall by definition is independent from the structural building frame, each component and the system as a whole must maintain structural integrity.

In this case, the curtain wall functioned well in resisting weather and maintaining the building's aesthetics, but the snap cover failures represented a lack of structural integrity.

Fortunately, most structural failures are serviceability related, not catastrophes, because engineers are trained by education and experience to design against failure. Because of the involvement of engineers, primary structural elements rarely fail, but secondary components like curtain walls are subject to a litany of failures because of the absence of their involvement.

The Remedy

Mapping the missing snap covers on elevation drawings was easy, as they were very conspicuous. The missing covers were then correlated with the shop drawings to discern a pattern.

The curtain wall shop drawings showed that hundreds of details were used, including numerous dies to extrude the various aluminum members. However, the snap covers only fell off four different sill members, and they shared a common die.

> ## The snap cover failures represented a lack of structural integrity.
>
> By David H. Nicastro

We concluded that the suspect die created extrusions with inadequate interference (the overlap dimension on two members that fit together). The interference is only a nominal dimension; the actual extrusions vary due to fabrication tolerances. Therefore, it was only necessary for this die to average toward the low range of interference, and statistically, a few would have inadequate bite.

While the low interference led to weak engagement of the snap covers to the sill extrusions, a force was still necessary to dislodge them—gravity and wind were not enough. We found two sources of loading that appeared to be significant:

1. Many of the snap covers that fell had holes drilled through them to allow them to pass over intermittent stabilizer anchors used with suspended scaffolding. The repeated action of connecting lanyards to the anchors (and subsequently disconnecting them) may have dislodged the weakly engaged snap covers.

2. Some of the snap covers may have been dislodged by slip-stick movement (sudden conversion of pent-up thermal stress into movement, rather than smooth, continuous thermal expansion and contraction).

As a remedy, we specified tugging every snap cover to test for engagement and securing the loose ones with high modulus (structural) silicone sealant used as an adhesive along the engagement lips. ◆

DAVID H. NICASTRO, P.E., founder and president of Engineering Diagnostics, Inc., in Houston, specializes in the investigation and remedy of construction problems and the resolution of related disputes.

Failure Mechanism: Slip-Stick Movement

Here is another case of things falling from a high-rise building. We often see falling concrete spalls and bricks, glass lites, cladding stones, and now, long aluminum extrusions. Water-induced failure mechanisms and leakage are the most common problems reported to us, followed closely by falling objects.

I have stayed in the hotel that is the subject of this case history, so I heard the slip-stick movement for myself. It can be frightening—a sudden, loud "bang," sometimes compared to a gunshot. Slip-stick movement refers to the nonuniform thermal expansion or contraction of a system, characterized by sudden movement ("slip") relieving accumulated ("stuck") stress. During construction, the restriction to sliding may have been caused by alkaline run-off water coming into contact with the aluminum members, such as from concrete placements above previously installed curtain wall components.

Structural Sealant Glazing

Since it was first used in commercial construction in the early 1970s, structural sealant glazing (SSG) has gained popularity for its ability to make curtain wall exteriors appear seamless. Instead of using aluminum extrusions with rubber gaskets to retain the glass in discreet windows, the glass appears to be continuous from lite to lite. This construction is possible because a high-strength silicone sealant can be used as an adhesive as well as a waterproofing material.

An infinite variety of details are possible using SSG, allowing architects great creativity. When properly designed and installed, these systems are virtually maintenance free. However, there is reason for alarm when the adhesive fails; nothing else holds the glass into the building.

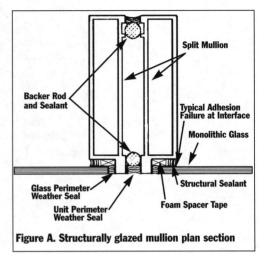

Figure A. Structurally glazed mullion plan section

When the adhesive fails, nothing else holds the glass into the building.

By David H. Nicastro

Assessing Seal Strength

The drawing is taken from a recent investigation of an 18-story building containing four-sided SSG. Two-sided SSG uses structural glazing on two sides of each glass lite (typically the vertical edges, to allow a continuous ribbon of glass) and conventional aluminum extrusions and gaskets on the other two sides (typically the head and sill). In four-sided SSG, all edges of the glass rely solely on a bead of silicone sealant for adhesion.

Most often, the structural bead is behind the glass and a separate weather-seal joint is used between the edges. In this case, a metal fin separates adjoining lites, so there are two weather-seal joints from aluminum to glass as well as the structural sealant joints behind the glass.

The owner initially reported water infiltration into tenant spaces during rainfall. We were immediately concerned with the structural behavior of the curtain wall because of the nature of the design. Leakage through a four-sided structural glazing system inherently implies that the structural seal retaining the glass has at least partially failed. Therefore, a catastrophic failure could ensue.

We were able to observe numerous adhesion failures at the structural sealant/glass interface simply by looking through the glass (which fortunately was not highly reflective in this building). However, we could not observe whether additional adhesion failures were occurring at the structural sealant/aluminum interface. Furthermore, the workmanship of the original installation was atrocious; the dimensions of these critical sealant beads varied widely, and voids were visible where the structural sealant had been incompletely installed.

The primary failure associated with SSG is fatigue. Repeated cycles of wind loading are expected to eventually cause adhesion failure. However, sufficient data is not yet available to predict reliably what service life remains in a system with intermittent failures, especially one with such widely varying deficiencies in original construction. Therefore, comprehensive remedial action was recommended.

Developing Standards

The American Society for Testing and Materials (ASTM), through its Committee C 24 on Building Seals and Sealants, is currently preparing guides for evaluating the existing condition of structural sealant glazing systems, including a test method for determining localized adhesion loss.

As the industry ages, it becomes increasingly important to develop standard protocols for determining the adequacy of these systems. While we are fairly confident that the silicone sealant will remain relatively unchanged for decades, the behavior of these systems also depends on the adhesion remaining adequate. That critical durability has yet to be determined. ◆

DAVID H. NICASTRO, P.E., founder and president of Engineering Diagnostics, Inc., headquartered in Houston, specializes in the investigation and remedy of construction problems and the resolution of related disputes.

Failure Mechanism: Adhesion Failure

This case disturbed me because of the number of different parties that did not understand the basic technology of structural sealant glazing (SSG).

- The glazer who installed the SSG with atrocious dimensional control and inferior adhesion.
- The curtain wall designer who must never have visited the job site during construction to observe the workmanship, which was uniformly inferior.
- The property manager who did not recognize that a leak in an SSG system is a symptom of a severe problem.
- A remedial glazer who replaced a broken glass lite with a lack of care similar to the original glazing.

We did not want to appear alarmist, but we were genuinely concerned about a potential catastrophe, and it seemed no one understood the severity of the problem. Perhaps SSG is not common enough for property managers to be aware of the hazard posed by adhesion failures, but certainly the other parties should have known.

In another case, I was called to investigate an SSG failure from a building that had numerous glass lites blow out one morning when the air-conditioning system was turned on, pressurizing the building. Again, there were early warning signs (and not subtle ones) of impending failure: water infiltration and copious exudate running down the glass from the structural sealant joints, which resembled maple syrup. This failure was caused by a chemical incompatibility between the structural sealant and a neoprene backing material. If such serious failures can occur after giving unheeded warnings, then I fear for public safety.

Zipper Gaskets

Last month, this column dealt with structural sealant glazing, a special type of exterior window system in which the glass is adhered to the metal frames with silicone sealant. Another system, structural gaskets (sometimes called "zipper" or "lockstrip" gaskets), is closely related but unique.

Both systems attempt to avoid disrupting the continuity of the glazing material on the exterior, providing an all-glass appearance (or glass and metal infill panels) without exterior mullions. Both systems are "barrier" designs, relying on the outermost components to be completely weatherproof, with no possibility of draining infiltrating water back to the exterior. And both systems place the heavy burden of structural performance (the ability to resist and transfer forces) on the exterior seals.

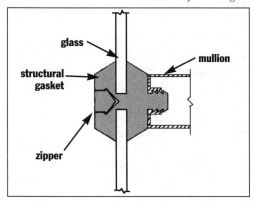

By David H. Nicastro

But there is one key difference: in my experience, properly constructed structural sealant glazing provides a durable, maintenance-free curtain wall, while structural gaskets never perform well.

Keeping Water Out

The drawing shows a typical structural gasket cross section of the head, jamb, and sill, which all look about the same. The rubber gaskets are intended to retain the glass in the opening, providing the only structural support for both gravity and wind loads. Generally, they perform this function adequately; loss of glass from structural gaskets is rare. Where the gaskets fail is in weatherproofing. These systems are notorious for leaking like sieves.

All gasketed window systems (even conventional roll-in gaskets with exterior aluminum extrusions) rely on the compression force of the rubber gasket against the glass to keep water out. This critical property is known as lip pressure and can be quantified.

Various devices have been manufactured to measure lip pressure, but they all are based on measuring the response of the gasket sample to deformation. No method has yet been devised to accurately measure the residual, in-situ pressure of a gasket against the glass.

But it is known that the lip pressure decreases over time due to shrinkage, compression set, and hardening as the gaskets age or weather. Gasket designers attempt to fabricate the rubber material properties, shape, and dimensions to maintain a compromise between minimum lip pressure to keep water out and the maximum that can be applied against the glass without breaking it.

In a conventionally glazed window, the gaskets are confined between the exterior aluminum extrusion and the glass and, therefore, can lose a great deal of lip pressure while still maintaining a seal between the two rigid materials. However, structural gaskets are not confined by exterior extrusions; the lip pressure is supposed to be maintained against the glass by a dense zipper strip (a separate rubber wedge that is harder than the gasket) inserted in the exterior of the gasket.

The design theory is that the zipper strip's internal force against the center of the gasket will be translated to the tips of the gasket against the glass. This mechanism does work, at least temporarily; but as structural gaskets lose resiliency, rebound from being stretched into place during construction, shrink due to off gassing of volatile plasticizers, and weather, they do not maintain an effective seal against the glass.

Water also typically enters structural gasket systems at the window corners (even initially), where the zipper strips are not as effective because of the geometry of the corners, and through butt joints in the gaskets.

Sealing the Seal

A range of remedies have been attempted, including replacing zippers, injecting sealants under the gasket lips, and wet-sealing the perimeters. I have found the only effective, long-term solution to be the application of a silicone sealant wet-seal bead around the perimeter of the lites, coupled with bridge joint detailing of the gasket butt joints, and filling the zipper channels with sealant before reinserting the zipper strips to prevent water from migrating through the channel to the gasket butt joints. ◆

DAVID H. NICASTRO, P.E., founder and president of Engineering Diagnostics, Inc., headquartered in Houston, specializes in the investigation and remedy of construction problems and the resolution of related disputes.

Failure Mechanisms: Inadequate Lip Pressure; Shrinkage;
Out-Gassing; Hardening

This column received more letters to the editor than all of the other columns *combined*; only a few were printed in subsequent issues. The letters were split between proponents of zipper gaskets who believed that I had maligned a good system, and those whose experience was similar to mine. As noted in the third paragraph, I was summarizing only my own experiences with zipper gaskets. There are no doubt many successful installations, but it is an occupational hazard that I see only the ones with problems.

I do not want to renew this controversy, so I will add only one comment. Although the column focused on design issues and material properties, it should be noted that there are at least as many (if not more) potential causes of failure of zipper gasket systems from improper installation.

The Devil is in the Details

Exterior insulation and finish systems (EIFS) recently have received an abundance of criticism because of a rash of failures throughout the United States. EIFS technology was imported from Europe; however, Europe has not experienced the problems common here. One key reason for the failures in the United States is the use of paper-faced gypsum sheathing as a substrate.

A Single Line of Defense

EIFS are non-load-bearing exterior wall cladding systems consisting of an insulation board adhesively or mechanically attached to a substrate, a base coat reinforced with fiberglass mesh, and a finish coat. While this general description is accurate for all systems, the exact nature of the individual components varies widely among the dozens of manufacturers.

Until recently, it was axiomatic that EIFS were barrier systems. Several manufacturers now offer system designs intended to accommodate some water penetration and behave more like a drainage wall. Still, most systems rely on a single line of defense against water infiltration—the base coat. The water resistance of the base coat is a function of thickness, aggregate size, composition, and porosity.

In the photo, a worker has knife-cut a rectangle in an EIFS and is peeling away the finish coat, base coat, and reinforcing mesh lamina, revealing the insulation board. These excavations were made to investigate the causes of severe distress in the EIFS (including pieces falling off) and water infiltration of these condominiums. Findings from this project were typical of most: While the base coat provides a dubious barrier to moisture through the face of the wall, most water penetration occurs at termination details.

In this case, the worst culprit was a poor roof eave detail that was dumping rainwater into the top of the wall. We also found points of water entry at sealant joints, window and door perimeters, penetrations, and the edges of balcony decks.

Once inside the wall, the water saturated the gypsum sheathing and migrated through the insulation board. In each of our excavations, we were able to remove the insulation board by hand because the paper face on the gypsum sheathing had deteriorated. The insulation board had been adhered to the sheathing (typical for this type of system) and not mechanically fastened. After this single finding, we recommended replacing the entire system because of the likelihood that large areas of the EIFS would fall off.

Under European codes, gypsum sheathing is not an allowable substrate. Because water penetration is likely in EIFS, gypsum sheathing (even water-resistant gypsum backing board) should not be used as a substrate for adhesively bonded EIFS.

Balcony railings (like the one visible to the right in the photo) presented another problem. The contractor had installed the railings by screwing them into the sidewalls of the balconies, which were constructed of EIFS. The insulation board and gypsum sheathing did not provide a suitable anchorage for the railings.

EIFS can work well when properly designed and constructed with high-quality materials, but the devil is in the details. ◆

> Europe has not had the problems with EIFS common in the United States.
>
> By David H. Nicastro

Courtesy the author

DAVID H. NICASTRO, P.E., founder and president of Engineering Diagnostics, Inc., headquartered in Houston, specializes in the investigation and remedy of construction problems and the resolution of related disputes.

Editor's Note

There are different schools of thought regarding EIFS. For the EIFS Industry Members Association's (EIMA) viewpoint, see Letters, page 8.

Failure Mechanisms: Deterioration; Cracking; Stress Concentrations

In every EIFS failure we have investigated, we found the primary cause of deterioration to be water absorption—but in no case did the water enter through the EIFS itself. Repeatedly, we find the same devilish details permitting water into the system: unsealed joints between EIFS and other materials, unsealed penetrations, and roof/wall intersections that do not shed water properly. (On this last point, see the discussion on page 41 regarding horizontal/vertical transition waterproofing.)

In addition to water-induced failure mechanisms, we have observed other common problems in EIFS construction.

- Cracking at points of stress concentration, such as at corners of openings, even when heavier mesh is properly installed.
- Cracking at nonstaggered joints—where the insulation board joints coincide with the sheathing joints.
- Incomplete mesh embedment in the base coat. It is acceptable to discern the pattern of the mesh in the base coat, but not its color.
- Inadequate fastening of the insulation boards to the sheathing.
- Non–back-wrapped edges of insulation boards, leaving an exposed edge without base coat and mesh.

For many years the EIFS industry was plagued with sealant joint failures. The leading contributor to the rash of joint problems was interphasal adhesion failure of the sealant to the finish coat; the finish coat would delaminate, leaving a thin layer of the acrylic finish on the cured sealant. The industry responded by recommending that sealant be adhered only to the base coat, not the finish coat. Apparently, this suggestion has worked well to reduce the initial joint failures, but a larger problem looms: how to effectively reseal a joint when the substrate is so soft that conventional tools would destroy the joint faces. The most common solution to this problem is to seal over the original joint with a bridge-joint profile, which is adhered to the finish coat, thus again raising the potential of interphasal adhesion failures.

My firm requires the use of primer on all sealant joints to promote adhesion, with one exception: when testing shows that the primer is deleterious. This caveat stems from the one time that I observed primer to lower the bond strength: the primer melted the finish coat on an EIFS installation.

Handrail Post Deformation

Many buildings are constructed with balcony handrail posts set into pockets or sleeves cast into a concrete deck or curb. Grout is used to secure the posts into the pockets. This simple detail is subject to extraordinary serviceability problems due to deterioration and difficult maintenance.

Ponding Water Damage

Balconies are exposed to substantial impinging rainfall and often to ponding water due to inadequate drainage. Handrail post pockets can collect water, leading to a host of problems:

• Corrosion of steel sleeves or posts. The vol-

Minimizing water collection at handrail posts is essential to maintaining durability.

By David H. Nicastro and Kimball J. Beasley

ume increase of corrosion can induce forces in the surrounding concrete, leading to cracking or spalling. The use of deicing salts and carpeting can exacerbate corrosion.

• Leakage into the perimeter walls or onto lower balconies, as the pockets often penetrate through the deck.

• Grout deterioration or erosion of non-water-resistant materials, which can lead to post loosening.

• Disintegration of epoxy or polyester resin grouts, which can shrink or crack. They may also thermally expand and contract differently than the adjacent concrete, causing internal stress.

• Ice crystal formation, which can cause deterioration of the grout or cracking or spalling of the surrounding concrete.

• Expansion of cementitious grouts containing gypsum as they become wet, leading to cracking or spalling of the surrounding concrete.

The handrail post in the photo exhibits a complete loss of support caused by expansion of the grout. In this case, the grout contained gypsum and expanded when exposed to moisture, pushing the post out of the pocket. We have also investigated cases of handrail post deformation that appeared similar but were caused by freezing water trapped in the hollow posts (ice crystal formation is a tremendous force of nature).

Preventive Measures

Concrete spalls at failed handrail posts can be a serious hazard. Immediate removal of any loose material is necessary once it has been discovered. If the source of the failure is expansion of an embedded element, such as a steel sleeve or gypsum-containing grout, it must be completely removed.

Courtesy Kimball J. Beasley of Wiss, Janney, Elstner Associates

Reanchoring of the existing handrails is sometimes possible, but in many cases the entire system must be replaced.

As with most problems, handrail post deformation is more easily prevented than remedied. Minimizing the collection of water at the handrail posts is essential to maintaining durability—note that all the failure mechanisms listed previously involve water. Balcony decks should be sloped to prevent ponding. Also, the posts should be mounted in a noncritical manner, such as on a raised curb or in stainless steel pockets. Finally, the handrail design should accommodate thermal expansion and contraction of long metal pieces without overstressing the connections to the structure. ◆

DAVID H. NICASTRO, P.E., founder and president of Engineering Diagnostics, Inc., headquartered in Houston, specializes in the investigation and remedy of construction problems and the resolution of related disputes. **KIMBALL J. BEASLEY, P.E.,** is a senior consultant with Wiss, Janney, Elstner Associates in Princeton, New Jersey.

Failure Mechanisms: Moisture Expansion; Ice Crystal Formation;
Swelling; Corrosion; Spalling

One statement near the end of this column stands out as remarkable to me:
". . . all the failure mechanisms listed previously involve water." Considering
how diverse these mechanisms are, it is phenomenal that they should all involve
water: corrosion, grout deterioration, erosion, ice crystal formation, spalling, and
swelling. Reviewing the list of failure mechanisms at the end of this book makes
it clear that water—as a vapor, liquid, or solid—is the real culprit behind an enormous number of mechanisms.

In addition to initiating material failures, water causes problems with buildings in a variety of ways.

- Leakage, which is annoying for tenants but also damages furniture, property, and interior finish materials.
- Loss of energy efficiency from wet insulation.
- Poor indoor air quality (and possible health effects) from micro-organism
  growth in damp, concealed spaces.

Sloped Glazing Slips

My colleague in Thailand, Peter Hartog of Building Diagnostics Asia Pacific, often writes and speaks about inadequate technology transfer between countries. Occasionally, the problem becomes apparent when construction techniques are used in the United States that were developed abroad. However, more often, he sees problems with attempting to adopt Western construction techniques in the Orient and, to a lesser degree, in

Do not tighten the mullions to stop leakage and sliding of the glass.

By David H. Nicastro

other regions. It appears that the problems stem from using basically sound design concepts without the benefit of local experience to fully understand proper detailing. Sloped glazing is one prevalent example.

Overcoming Gravity

The photograph shows a glass awning over the sidewalk around a shopping mall in Sydney, Australia. The glass is slowly slipping down the slope under the influence of gravity; gaps up to 75 mm can be seen along the top of the glass. Eventually, the glass rests against a steel gutter at the bottom and fractures due to edge contact. Fortunately, the lites are composed of laminated glass.

The design of this system relied on the adhesion of glazing tape and the clamping pressure of the mullions to resist gravity and wind loads, but it did not use bottom-edge stops. Differential thermal movement caused the adhesion of the tape to fail, allowing gravity slowly to overcome the friction.

As is common in sloped glazing, this installation used silicone sealant in the transverse butt joints to avoid a metal mullion that would interrupt free drainage of rain water. The silicone joints are strong enough to remain adhered even as the glass slides, so both the upper and lower lites are sliding together.

Courtesy Peter Hartog

While the silicone is performing its intended function well, it has contributed to another failure with this system: fogging of the laminated glass lites. Loss of visual clarity can be seen at all but one of the butt joints, due to chemical incompatibility of the sealant and the internal PVB film used to laminate the glass. PVB-laminated glass is also susceptible to fogging from contact with moisture; this additional failure mechanism can be observed along the edges of the lites.

Arresting Failure

A common early warning sign of failure of skylights and sloped glazing is water infiltration. In similar cases, we have seen contractors futilely attempt to arrest leakage and sliding by tightening the mullions—sometimes breaking the glass. The simple addition of bottom edge stops would remedy the displacement problem, but only reglazing can resolve the widespread fogging. Incidental leakage can be addressed by wet-sealing, unless the leakage indicates a systemic failure of the perimeter restraint system. ◆

Acknowledgment

This article was prepared in conjunction with material provided by Mr. Hartog.

DAVID H. NICASTRO, P.E., founder and president of Engineering Diagnostics, Inc., headquartered in Houston, specializes in the investigation and remedy of construction problems and the resolution of related disputes.

Failure Mechanisms: Fogging; Edge Contact; Fracture;
Thermal Expansion

One problem with the transfer of technological information, between countries or even domestically, is lore: the accumulated conventional wisdom, which may be factually incorrect and is often outdated. As an example, delamination of laminated glass from incompatible sealant contact was generally believed, for many years, to be caused solely by acetoxy-curing silicone (the kind that smells like vinegar). A major manufacturer of polyvinyl butyral (PVB) film used in laminated glazing, however, found that the most extensive delamination was caused by a neutral-curing silicone.

Unfortunately, because of the widespread belief that neutral-curing silicone would not cause delamination, it has been used in contact with laminated glass throughout the world, resulting in countless failures. It is now nearly impossible to overcome the erroneous conventional wisdom and educate the construction industry on the more recent recommendations. The communication problem is magnified between geographical regions, with different languages and cultures.

Fenestration Frustration

According to numerous surveys, water infiltration is the source of 50 to 75 percent of construction litigation claims. While this is a staggering figure, it is easy to understand: water causes not only leaks and interior finish damage but also mildew and poor indoor air quality, loss of insulating value of soaked materials, and a myriad of other failures, including decay and corrosion.

Appearances Can Be Deceiving

Most often, water infiltration through exterior walls appears as window leaks. The windows do leak in many cases, but more often they simply create a penetration that interrupts the exterior waterproofing systems. Hollow metal frames also act as conduits, channeling water from remote locations to the window sills or heads.

Windows create penetrations that interrupt exterior waterproofing systems.

By David H. Nicastro and Kenneth B. Simons

In residential and light commercial construction, self-flashing window units are common. These prefabricated frames come complete with a metal flange around the perimeter for nailing and can be used as a flashing tie-in to the dampproofing system. Designers often indicate that the windows should be installed according to the manufacturer's instructions, but the manufacturers usually do not provide instructions for the waterproofing tie-in. The result is an epidemic of improper installations.

Proper Installation

The figure shows the proper installation of a self-flashing window unit. The flanges should be pressed into a continuous bead of sealant before nailing. Then 230 mm (9 in.) flashing strips are installed over the flanges, and the underlayment (such as building paper) is installed in a shingled (overlapping) pattern. Finally, siding or masonry is installed as a veneer.

When these window units are not properly installed and leakage develops around the window, the tendency is to apply copious amounts of sealant over all possible external openings. This is unfortunate because the design intent of these units (if it were clearly specified by the designer or manufacturer) is to provide multiple lines of defense against water infiltration. The flanges, flashing, and underlayment act as the last barrier, but they must weep collected water through the veneer someplace; inappropriate sealant application can block the weep holes.

If removal and proper reinstallation are not possible, then remedial sealant applications must be carefully designed to reduce the amount of water being handled by the internal flashings while permitting an escape route for any water that does migrate through the first line of defense. ◆

DAVID H. NICASTRO, P.E., founder and president of Engineering Diagnostics, Inc., headquartered in Houston, specializes in the investigation and remedy of construction problems and the resolution of related disputes. **KENNETH B. SIMONS, PH.D., P.E.,** principal engineer for Damage Consultants, Inc. (DCI), in Bellevue, Washington, specializes in the investigation of various types of damage to constructed facilities, problem solving, troubleshooting, and repair design.

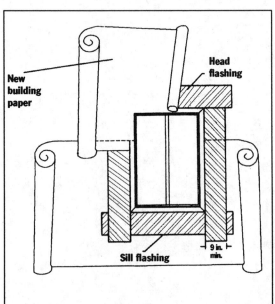

Figure A. Exterior window and door penetration flashing (scale: NTS)

Courtesy Kenneth B. Simons

Failure Mechanism: Not Applicable

No industry standard has been established for the proper flashing of windows. The Uniform Building Code simply requires that "exterior openings exposed to the weather shall be flashed in such a manner as to make them weatherproof." Therefore, there are a variety of methods in different regions of the country. This column, however, presented a method that is not appropriate.

In attempting to edit this column to fit the available space, the description of the proper flashing installation was inadvertently changed. The correct installation procedure requires that the jamb and sill flashing strips be installed beneath the flanges and that the flanges be pressed into a continuous bead of sealant before nailing. Next, the head flashing strip is installed, and the underlayment (such as building paper) is installed in a shingled (overlapping) pattern. The following drawings should further clarify the method.

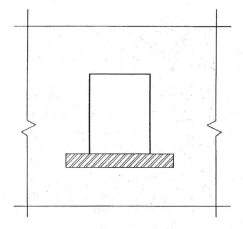

STEP 1. Attach Sill Strips Level with Sill and Extend out beyond Opening.

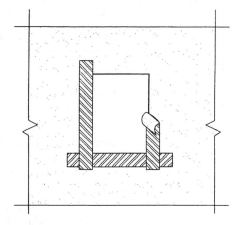

STEP 2. Attach Jamb Strips Even with Opening Sides and Extend above Head and below Sill Strip.

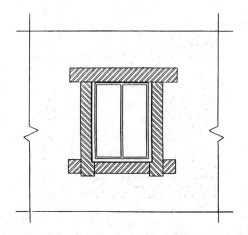

STEP 3. Place Window into Opening with Nailing Flanges over and Sealed to the Sill and Jamb Strips with a Compatible Sealant, then Install Head Strip.

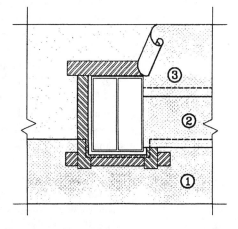

STEP 4. Install Underlayment Beginning at the Bottom with Layer 1, such that It Lies Beneath the Sill Strip and Continuing with the Additional Overlapping Layers 2 and 3, up the Wall, over the Jamb and Head Strips.

The Rotating Beam

Designers sometimes use "floating walls"— masonry walls that project beyond the building's perimeter column line above the first floor level— to create an open, uninterrupted first floor and a dramatic visual effect. Unfortunately, many designers are not accustomed to dealing with the unusual loading conditions this construction imposes, and they overlook the spandrel beam's torsion capacity. The result is a distorted beam and cracked and displaced masonry.

Many designers are not used to floating wall loads.

By David H. Nicastro and Kimball J. Beasley

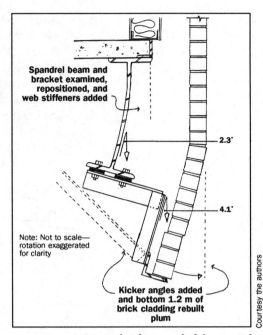

Spandrel beam and bracket examined, repositioned, and web stiffeners added

2.3°

4.1°

Note: Not to scale— rotation exaggerated for clarity

Kicker angles added and bottom 1.2 m of brick cladding rebuilt plum

Courtesy the authors

Structural Remediation

Floating walls are usually supported by a perimeter spandrel beam that is connected to transverse beams cantilevered from the columns. In addition to normal bending forces, the eccentric masonry's dead load subjects the spandrel beam to rotational forces (torsion).

A large retail store in the Northeast was recently constructed with 11 m (35 ft) tall exterior brick veneer walls, projecting about 0.6 m (2 ft) beyond the perimeter column line above the second floor level. Rotation of the second floor spandrel beam and a corresponding inward displacement of the bottom of the floating wall was not noticed until the wall was completely built.

A subsequent investigation revealed that the projecting brick wall was supported by a steel shelf angle hung from welded angle and "T" brackets, which were bolted to the bottom flange of the spandrel I-beam. No spandrel beam web stiffeners or lateral spandrel beam supports were provided in the original design. Calculations indicated that the rotational forces from the brick dead load stressed the spandrel web to near yield. To complicate the situation, few wall ties had been used along the bottom 1.2 m (4 ft) of brick cladding because it was difficult to install ties at the spandrel beam.

The investigators concluded that the masonry could collapse because of the displacement, the lack of wall ties, and the overstressed I-beam web. Therefore, major remedial structural work was implemented. The top 9 m (30 ft) of the wall was temporarily shored and the bottom 1.2 m of masonry was removed. The spandrel beam and brackets were examined for cracking and inelastic deformations. The existing brackets were repositioned, new steel angle kicker braces were installed, and new spandrel web stiffeners were welded to the exterior side of the spandrel beam. The bottom 1.2 m of masonry was then rebuilt plumb with customized wall ties.

Failures During Construction

While this retrofit work was being performed, the area around the exterior walls was restricted, but the building remained open. Unlike new construction, it is often necessary to perform remedial work while a building is open to the public. Although comprehensive data is not available, various studies have indicated that up to 70 percent of all failures occur *during* construction (including remedial construction); therefore, designers are cautioned to take this additional risk to the public into account when planning remedial work. ◆

DAVID H. NICASTRO, P.E., founder and president of Engineering Diagnostics, Inc., headquartered in Houston, specializes in the investigation and remedy of construction problems and the resolution of related disputes. **KIMBALL J. BEASLEY, P.E.,** is a senior consultant with Wiss, Janney, Elstner Associates in Princeton, New Jersey.

Failure Mechanisms: Eccentric Load; Cracking

The discussion at the end of this column regarding failures during construction was prompted by the collapse of a portion of the Northline Mall in Houston, which occurred while I was writing the column. Remedial construction was being performed on a portion of the mall when a wall collapsed, killing three shoppers.

Not only is construction a risky business for the contractor's personnel, but it imposes hazards to the public when performed on an occupied building. Although several of these columns boast of not closing the building to the public during remedial construction, we must also be aware of the risk to the public that was imposed for the owner's convenience.

CATALOG OF FAILURE MECHANISMS:
CAUSES AND CHARACTERISTICS

This catalog of failure mechanisms is presented in alphabetical order, with primary entries based on the most common names of recognized failure mechanisms. Where alternate names were found, they are included either as separate main entry headings with a reference to the preferred term, or they are listed after a comma with the main entry heading if the alternate terms were considered to be of nearly equal importance. Capitalized terms in the body of an entry refer to additional entries.

Although most of the failure mechanisms listed are processes, some are simply original construction defects. Consider "honeycombing": although this deleterious concrete condition is not truly a mechanism, it meets the working definition of a failure mechanism because it can be identified as the technical cause of a failure.

Abrasion. The wear or removal of the surface of a solid material as a result of relative movement of other, usually harder, solid bodies in contact with it.

Adhesion Failure. A failure in an adhesive bond characterized by visible separations between two formerly bonded, intact materials, or characterized by such a weak bond that it fails with very little applied force (an incipient Adhesion Failure). Most Adhesion Failures occur at the interface between the two materials (interfacial Adhesion Failure); an Adhesion Failure can also be interphasal: immediately adjacent to the bond line within one of the materials. Interfacial Adhesion Failures can be caused by adhering a material to a wet substrate, frost, dirt, contamination, Corrosion, or a degraded substrate; lack of priming or primer saturation; chemical incompatibility with substrate; or Overstress. Interphasal Adhesion Failures are usually associated with water saturation and may be caused by a weakening of a material when wet, which Delaminates under applied load.

Aging. An imprecise term for time-dependent, inherent Deterioration or a change in properties. Aging should not be used as a synonym for Weathering, which is similar but also involves external actions by the environment.

Algae. Any of a group of nonvascular plants. As an algae colony grows and becomes more established, its roots can destroy building materials.

Alkali-Carbonate Reaction. In concrete, the reaction between alkalies (sodium and potassium) in portland cement and certain carboniferous stone or minerals used as aggregate, such as calcitic dolomites and dolomitic limestones. Products of the reaction may cause expansion and cracking of concrete in service. The reaction releases ions that migrate with water into the restricted space of the fine-grained matrix surrounding the aggregate. The growth and rearrangement of the crystals exert internal pressure (American Society 1990).

Alkali-Silica Reaction. The reaction between alkalies (sodium and potassium) in portland cement (and rarely from admixtures, de-icing salts, or other aggregates) and certain siliceous rocks or minerals present in some aggregates, including opal when greater than 0.5% of the aggregate; chalcedony as a constituent of chert that is greater than 3%; and acidic volcanic glass. The reaction forms a gel around the aggregate, which absorbs water and swells. The internal expansive forces may cause cracking of concrete in service. The cracks tend to have a longitudinal orientation. Extremely reactive materials can cause cracks within a year; slower reactions are more typical, with cracks appearing after 20 yr. Once cracking commences, secondary failure mechanisms are exacerbated. Moisture is necessary for the gel to form and expand (American Society 1990; Pryor 1992; Swamy 1992).

Alligatoring, Alligator Cracking. In built-up roofing, a distinct pattern of shrinkage cracks in bitumen. The cracks may not extend through the surface bitumen. In other materials, see the preferred term Crazing.

Attack. A deleterious chemical Reaction that harms a building material.

Bitumen Floating. In built-up roofing, the time-dependent upward migration of inter-ply bitumen to the membrane surface under the action of heat and load. This process weakens the roof membrane as the felts become dry and unbonded.

Bleeding. In concrete, the separation of water from an unhardened mix; the escape of excess water during curing. Bleeding can cause Water Pockets. In paints, the process of diffusion of a soluble colored substance from, into, and through a paint or varnish coating from beneath, thus producing an undesirable Staining or Discoloration. Also, the transfer of soluble material from bitumen-impregnated roofing materials, in lime-rich water, causing Staining of soffits of concrete slab roofs (American Society 1990).

Blister. A local separation or Delamination between two layers of a coating or membrane, or between the substrate and the coating or membrane, characterized by a raised area on the surface with a cavity below, sometimes filled with liquid or gas. Blisters are usually caused by penetration of moisture through areas of poor adhesion (Tam and Stiemer 1996).

Bloom. The accumulation of residual fluid components on the surface of a sealant or coating. The fluid can be high- or low-molecular weight; it is a common misconception that only the lower molecular weight fluids can migrate within a material. Another misconception is that the accumulating fluid is always plasticizer. Plasticizers are only one component of a product and not the only fluid that can migrate. Bloom is also sometimes used as a synonym for Efflorescence, the preferred term.

Bond Failure. The slippage of steel reinforcing bars in the surrounding concrete. In other materials, see the preferred term, Adhesion Failure.

Breadloafing. Swelling of an organic elastomeric sealant, usually associated with early water contact before complete curing.

Brick Suction. The absorption of moisture by Capillary Action into brick units produces a suction effect that draws water out of the fresh mortar or grout. Low initial rate of absorbtion (IRA) units (<5 g/30 in.^{2}/min) do not absorb much water; the units may float on the mortar, thus producing a poor bond. High IRA units may absorb too much water from the mortar, causing the mortar to stiffen too rapidly. Empirically, the ideal IRA appears to be below 30 g/30 in.^{2}/min. Other factors that affect the mortar/brick bond strength include mortar Shrinkage and the moisture retentivity (resistance to brick suction) of mortar, which varies with its component proportions (Drysdale et al. 1994; Brick 1994).

Brittle Fracture. Structural steel subjected to tensile stresses and low temperatures may fracture by cleavage below the yield stress, with little or no plastic deformation. Brittle Fractures are often associated with triaxial tensile stresses, increased strain rates, cold work, residual tensile stresses from welding, thick plates, and notches (Stress Concentrations) (American Institute 1986).

Bubbling. In elastomeric sealants, the development of (and subsequent detrimental rupturing of) bubbles within the sealant body. Bubbling is usually caused by gas escaping from the sealant, backer material, or substrate, generally from one of the following mechanisms:

1. Sealant installed over a wet substrate or frost; the water subsequently boils off. This is especially prevalent on highly porous substrates.
2. Out-Gassing of some backer materials if punctured during installation and if they contained trapped gas from manufacturing.
3. Excessive air voids in the sealant beads, possibly entrapped during mixing of two-part sealants or improper tooling.
4. Applying sealant to a hot substrate, causing a quick surface skin to form before gases emitted during curing can escape.

Buckling. A sudden deformation or loss of straightness normal to the direction of applied loads in a structural member subject to compressive stress (or in the compressive zone of a member subjected to bending). See also Euler Buckling, Flexural Torsional Buckling, Lateral Torsional Buckling, Local Buckling, Oil-Canning, Torsional Buckling, Web Buckling, and Web Crippling.

Capillarity, Capillary Action. Absorption of a liquid due to surface tension forces through narrow openings. The distance that water will move through capillary tubes is a function of the diameter of the tubes and temperature.

Capillary Rise. Upward absorption of a liquid, usually water, into porous building materials, caused by Capillarity.

Carbonation. In concrete and other cementitious materials, the transformation of the free alkali and alkali-earth hydroxides in the cement matrix into carbonates, due to reaction with carbon dioxide in the atmosphere. This process occurs slowly, with the rate dependent on the porosity of the concrete and humidity. Carbonation is usually associated with moderate humidity (50–80% rh). Some moisture film is necessary in the pores to dissolve the hydroxides, but excess moisture prevents gas diffusion. Carbonation is not reversible and causes latent Shrinkage of concrete and concrete masonry units. It also lowers the passivating alkalinity of concrete, thereby accelerating Corrosion of embedded steel reinforcing. Carbonation occurs inward from surfaces exposed to air, rarely exceeding a depth of 30 mm, but can reach greater depths at Cracks and poorly consolidated areas (Murdock et al. 1991; Rosenberg et al. 1989).

Cavitation Corrosion. A form of Pitting Corrosion resulting from the formation and collapse of vapor bubbles in a liquid near the surface of a metal. The collapse of the vapor bubbles generates shock waves that produce very high local pressures that cause plastic deformation of the metal, a roughened surface, and many closely spaced pits. The pits permit the ingress of corrosive agents (Von Fraunhofer 1974).

Cement-Aggregate Reaction. In concrete, any Reaction between cement and aggregate, including Alkali-Silica Reaction and Alkali-Carbonate Reaction, but usually implying another Reaction such as hydration of anhydrous sulfates, rehydration of zeolites, wetting of clays, and reactions involving solubility, oxidation, sulfates, and sulfides. The products of these internal Reactions may cause expansion and cracking of concrete in service. These Reactions require the presence of water (American Society 1990).

Chalking. The formation of a powder (chalk) on the surface of a polymer-based material caused by the disintegration of the polymer component under the action of Weathering. Chalk can be any color but is most often whitish (Grassie and Scott 1984; Tam and Stiemer 1996).

Checking. In wood, short axial separations of fibers along the grain caused by the nonuniform loss of moisture; the outer wood dries faster than the inner core, producing a tensile stress that exceeds the capacity of the member, resulting in cracking and a loss of shear strength. In paint, a synonym for Crazing. In other materials, see the preferred term Crazing (Breyer 1993).

Chemical Weathering. Attack of limestones, dolomites, and marbles by sulfurous and sulfuric acids (such as from acid rain), and to a lesser extent by carbonic acid and ammonium salts. The Reaction of the sulfur-based acids and the stone form gypsum, whereas carbonic acid and ammonium salts dissolve the lime component, resulting in loss of stone. The cyclic process of dissolution and precipitation can increase porosity and lead to Cracking (Chin et al. 1990).

Chloride Attack. Attack of concrete by chlorides of ammonium, magnesium, aluminum, and iron: those of ammonium are the most harmful. Sodium chloride is chemically harmless to portland cement concrete but is a major contributor to Corrosion of embedded reinforcing steel (see Chloride-Accelerated Corrosion) (American Society 1990).

Chloride-Accelerated Corrosion. Corrosion of ferrous metals accelerated by chloride ions, which act as catalysts. This mechanism is especially prevalent in reinforced concrete. The steel iron atoms release two electrons to form a positively charged iron ion, which combines with an oxide ion to form a natural, passive iron oxide barrier on the steel. Chloride ions react with the positively charged iron ions to form a product that has the capability to override the passive layer. Inside the passive layer, the iron ions discharge the chloride ions, allowing them to cycle. At the cathode region, the free electrons react with water and oxygen to form hydroxyl ions, which cause destruction of the passive layer. At the anode region, the iron reacts with the hydroxyl-forming iron oxide, which migrates into the adjacent voids and reacts with additional oxygen to form expansive products (Abo-Qaidis and Qadi 1995).

Cissing. The recession of a wet paint film from small areas of the surface, leaving either no coating or an uneven one.

Cohesion Failure. The loss of integrity of a formerly intact material characterized by the development of fractures or separations within its body. It is generally implied that Cohesion Failures are full depth (completely through the material), but they may be intermittent along the length of the material and not continuous. Although occasionally caused by inadequate tensile strength of the material, cohesive failures are more often associated with Overstressing, including movement during cure (before the material attains its final properties to resist tearing).

Cold Joint. In concrete, a nonintegral interface and, therefore, a plane of weakness caused by placing fresh concrete against a pour that has already set.

Color Change. Any change (not necessarily adverse) in the color of a building material over time. If used to imply an unintended change in color, see the preferred term, Discolor.

Concentration Cells Corrosion. Galvanic cells in which the electromotive force (emf) is due to the difference in the concentration of one or more reactive constituents of an electrolyte solution. If metal is exposed to a nonhomogenous environment, then anodic and cathodic areas can form and cause Corrosion. Cells usually become concentrated in stagnant areas such as lap joints, crevices, or under fastener heads. The two most common types of cells are the following:

1. Metal-ions. Metal-ion cells dissolve at a slower rate when the concentration of its ions increases in solution; metal that is in contact with the more dilute solution will become more anodic, and metal in contact with the more concentrated solution will become more cathodic.
2. Oxygen cells. Low areas of oxygen can become relatively anodic, whereas a concentration of oxygen cells can become cathodic (Pludek 1977; Bosich 1970).

Concrete Corrosion. An imprecise term for Deterioration of concrete. It may occur in conjunction with Corrosion of embedded steel. See the preferred terms Attack, Chloride Attack, Corrosion, and Deterioration.

Condensate Corrosion. Localized Corrosion that occurs where moisture-laden hot gases contact a cooler metal surface.

Condensation. The precipitation of liquid from a vapor resulting from lowering the temperature at constant pressure, especially the deposition of water from warm moist air onto a relatively cold surface. As the condensate forms, the partial pressure of the gas adjacent to the condensate is lowered, resulting in diffusion from more remote (and higher partial pressure) regions. Therefore, an engine devel-

ops that allows Condensation to continue. Condensation of water can be a problem in and of itself, but most often is a contributing cause to other more significant failure mechanisms (such as Corrosion, Condensate Corrosion, Decay, Staining, and Mildew) because it provides a source of liquid water.

Cone Failure. Pull-out of the cone of concrete that resists axial tension on an embedded fastener (such as a headed stud or an undercut expansion anchor). The ratio of the concrete surface to the effective embedment depth is about 3:1, and the included angle is about 35°. Close spacing of adjacent anchors may cause truncated cones (Michaud 1996).

Connection Loosening. In wood, the time-dependent loosening of fasteners due to Shrinkage of the wood. The degree of Shrinkage, and therefore of the loosening, is directly related to the moisture content of the wood.

Consolidation. A decrease in volume of saturated cohesive soil as water is expelled from the pore spaces under load. Consolidation is time dependent, with the rate depending on the porosity of the soil (the rate of pore water drainage) and the amount of water present (Das 1994; Hóltz and Kovacks 1981).

Consumption. The destruction, radical conversion, or disappearance of a building material by any one of a variety of processes or organisms that consume the material, including fire, termites, beetles, worms, enzymes, rodents, fungi, Microbial Corrosion, Decay, and bacterial fermentation. The term is also used to describe severe damage and loss of strength of concrete caused by high heat, such as exposure to fire.

Contaminated Aggregate. The presence in or on aggregate used in concrete of organic impurities that interfere with the chemical reactions of hydration.

Contaminated Mixing Water. The presence in water used in concrete of organic impurities or chlorides that interfere with the chemical reactions of hydration, cause Staining, or cause Corrosion of the reinforcing steel.

Corrosion. The deterioration of a metal by an electro-chemical Reaction with its environment or another material with which it is in contact. Occasionally used imprecisely to apply to nonmetallic materials, for example, concrete and masonry (Bradford 1993).

Corrosion Fatigue. Development of cracks in metal caused by the combined effects of Corrosion and Fatigue, which interact to be more destructive than the sum of the two processes acting individually. For example, Corrosion can reduce the fatigue resistance of a metal from 50,000 to 5,000 cycles (Pludek 1977; Bosich 1970).

Corrosion Undercutting. The development of surface Corrosion on a substrate beneath a coating. Corrosion Undercutting is usually associated with water penetration through non-intact or nonadhered coating areas (Tam and Stiemer 1996).

Crack, Cracking. The separation of a material into parts and the resulting manifestation of the separation; an essentially linear discontinuity produced by fracture; or an elongated, narrow opening. Cracks are characterized by length, direction, spacing, width, depth, shape, activity, faulting, and pattern. Synonyms in various applications can include break, split, fracture, fissure, separation, and cleavage (American Society 1990).

Crazing. A network of cracks in a surface that do not penetrate through the material. Crazing is used to generally describe surface cracking of masonry, coatings, sealants, gaskets, and membranes. Crazing also specifically describes the fine network cracking of ceramic glazes by, for example, differential Thermal Expansion between the glaze and tile body, or moisture expansion of the body. For most organic materials, crazing represents normal, inevitable weather-induced Deterioration. When this process begins early in service, however, it may indicate a low-quality product. Crazing in organic polymer materials is a chain scission mechanism: the long polymer molecules can be severed by incident radiation or oxidation. As surface cracks form and expose fresh material from the interior to Weathering (especially ultraviolet radiation or ozone), scission can progress, eventually developing into Cohesive Failure or a Crack. The oxidation mechanism involves the development of surface hardness, which cracks, exposing fresh material to oxidation.

Creep. Time-dependent plastic flow (increase in strain under essentially constant load). In wood creep is dependent on moisture content and temperature. In concrete Creep is associated with the plastic flow of the paste, not the aggregate. Factors influencing the amount of Creep in concrete include: the duration of curing before load application; the compressive strength of the concrete; temperature; humidity; water-cement ratio; and the amount of reinforcing. Under high loads, Creep in unit masonry can become significant because of plastic flow of the mortar. Creep in concrete masonry units (CMUs) is approximately the same order of magnitude as ordinary concrete but typically occurs after the first year, whereas most Creep in concrete occurs in the first year. In addition to the factors listed for concrete, the factors influencing the amount of Creep in CMUs include the curing method and the density of aggregate. The Creep in the mortar joints of CMU masonry is often four to five times the Creep of the CMUs themselves (American Society 1990; Drysdale 1994; McCormac 1993; Breyer 1993).

Crevice Corrosion. A localized form of Concentration Cells Corrosion, which develops between metals in contact or between metals and nonmetals. Typical crevices susceptible to Corrosion include: tight joints that are just loose enough to allow moisture to penetrate, such as threads of bolts or lap joints; metal contacting an absorbent material, such as gaskets, insulation, coatings, or wood; and metal surfaces where deposits have formed, such as bacteria colonies or salts (Bradford 1993).

Cryptoflorescence. A salt accumulation inside masonry pores that is deposited as salt-laden moisture in the masonry evaporates, sometimes resulting in Staining. The opaque salt deposits near the surface of the material alter the way light is reflected and refracted from the surface, resulting in aesthetic detraction (American Society 1990).

Crystal Packing. In glass fiber-reinforced concrete (GFRC), the dominant mechanism that causes the material to lose ductility. Calcium hydroxide and calcium silicate hydrates form around and within the bundles of glass filaments. The process begins as these solutions are drawn into the fiber bundles by capillary action, and later crystallizes, filling the pores and "locking" the reinforcing fibers into the cement matrix. This prevents the fiber slippage, which creates ductility. After Aging (perhaps 10 yr), the ductility can approach that of an unreinforced panel. Because the process begins with an aqueous solution, preventing liquid water migration within the panels (such as with a thermoplastic copolymer curing additive) can theoretically prevent Crystal Packing. Other research is focusing on the addition of pozzolanic materials and metakaolin (McDougle 1995).

D-Cracking. The progressive formation on a concrete surface of a series of fine Cracks at rather close intervals, often of random patterns, but, in highway slabs, paralleling edges, joints, and Cracks and usually curving across slab corners. D-Cracking is often associated with dark-colored deposits or Staining that surrounds the cracked area and is commonly caused by Freeze-Thaw expansion fracturing of the coarse aggregate. This process usually starts at the bottom of a slab with approximately horizontal cracks, which become more random as they progress upward (American Society 1990).

Dealloying. A form of Selective Attack in which certain atoms are leached out.

Debond. See the preferred term, Disbond.

Decay. Any of a number of organic processes in which organisms Consume the cellulose in wood or paper building materials. The organisms typically require sufficient oxygen, a moisture content over 20%, and temperatures between 40 and 105°F. Decay mechanisms include Dry Rot, Wet Rot, Soft Rot, and Mold (Singh and White 1997).

Decomposition. The separation of a material into elements or parts, usually by organic processes such as Decay.

Dedolomitization. In marble the change of dolomite into calcite through Weathering as magnesium atoms in dolomite crystals are slowly replaced by calcium atoms (Cohen and Monteiro 1991).

Degradation. The lowering of a material's characteristics (such as strength or integrity). Similar to Decomposition but not necessarily implying an organic process. Also similar to Deterioration but not necessarily time-dependent.

Delamination. The breakdown of a material by its separation into layers. Nonintegral materials can separate along original lamination planes (adhesive), and both integral and nonintegral materials can develop internal separation planes (cohesive). In concrete, delamination refers to horizontal splitting, cracking, or separation of a concrete member in a plane roughly parallel to and generally near the surface. It is found most frequently in slabs and caused by the Corrosion of reinforcing steel or Freeze-Thaw and is similar to Spalling, Scaling, or Peeling except that Delamination affects large areas and is not readily apparent without sonic testing. Also, it is a condition of stone in which the outer surface splits apart into laminae or thin layers and peels off the face of the stone (see the preferred term Exfoliation). In coatings an Adhesion Failure between coats or between the substrate and the coating caused by Chalking, contamination, or an over-cured surface. In laminated glass an Adhesion Failure of the polyvinyl butyral (PVB) interlayer (American Society 1990; Tam and Stiemer 1996).

Delayed Ettringite Formation. In concrete latent development of ettringite crystals in the paste surrounding aggregate, caused by overheating of the concrete during curing or excessive sulfate in the cement. Early ettringite formation and fine ettringite needles in air voids and open cracks in old concrete are considered harmless, but distress can ensue from ettringite formation in hardened concrete. Large ettringite formations cause paste expansion and consequent gaps between the aggregate and paste, and can lead to concrete cracking in a pattern similar to that caused by Alkali-Silica Reaction. There are two recognized theories that attempt to explain the expansion: (1) The ettringite crystals grow within the cement paste until completely filled and then push apart the paste; and (2) The ettringite crystals grow until the air void volume is completely filled and then swell upon absorption of water (Marusin 1994; Lawrence 1995).

Deposit Corrosion. A form of Concentration Cells Corrosion caused by moisture retained in deposits on a metal surface, including deposits of previous Corrosion product (Mattsson 1989).

Deterioration. The gradual adverse loss of physical or chemical properties of a material (American Society 1990; American Concrete 1979).

Diagonal Cracking. Cracks resulting from high diagonal tensile stresses that develop along planes perpendicular to the principal tensile stress.

Diagonal Tension Failure. Excessive diagonal principal tensile stress in a structural member, usually a beam. Diagonal Tension Failures are often brittle and sudden but may be preceded by Diagonal Cracking (McCormac 1993; Nawy 1996).

Dielectric Corrosion. See the preferred term Galvanic Corrosion.

Differential Movement. Nonuniform movement between two components or two regions of the same component, either in opposite directions or of different magnitudes in the same direction.

Differential Settlement. A form of Differential Movement caused specifically by Settlement of supporting soil.

Differential Support. Varying degree of support between two regions of an integral component or system, which induces stress. Differential Support may be provided by materials with different moduli of elasticity or by discontinuities in the supporting elements.

Dirt Pickup. Soiling caused by a foreign material (other than micro-organism growth) deposited on or embedded into a surface. When Bloom is caused by high molecular weight components, dirt can become trapped in the fluid; this is a common cause of Dirt Pickup. Other causes include wind-induced electrostatic charge, which attracts dust and lightweight particles; tackiness; and surface profile (roughness).

Disbond. To suffer an Adhesion Failure.

Discoloration. An unintended change in the color of a building material over time, usually implying a process acting on the material itself, which cannot be reversed or cleaned off, as opposed to a superimposed foreign material (see Staining). The most common cause of Discoloration is the Weathering of pigments. Some pigments simply have better radiation resistance and color stability than others. Discoloration, however, may also indicate a chemical Reaction with another material with which it is in contact.

Discontinuity. An internal or external localized area of interrupted filler material in welded joints that results from cracking, entrapped slag, inclusions, deoxidation products, gas pockets, or blow holes (American Society 1990).

Disintegration. The loss of integrity or cohesiveness of a material, reducing an intact member into small fragments or particles. Disintegration can be sudden (due to factors such as Impact) or gradual (due to factors such as Attack, Weathering, Freeze-Thaw, or Erosion) (American Society 1990).

Downdrag. The unexpected additional load on a friction pile caused by the soil (usually fill) settling more than the pile. Downdrag can be caused by soil Shrinkage from desiccation or a receding ground-water table, and can result in a pile pulling out of its pile cap or the pile cap pulling free of the supported structure.

Dragdown. During pile driving, dragging down the soil (usually the top layer) with the pile. Dragdown can cause an increase in pile capacity in sandy soils or a decrease in capacity in soft clays (Tomlinson 1971).

Drift. Lateral deflection of an entire building (especially multistory), usually due to wind or seismic loading; also, the difference in horizontal deflection at the top and bottom of one story (also called story drift) (LRFD 1986; McCormac 1989).

Dry Rot. Decay of wood caused by a fungus that leaves the wood lightweight, friable, and dull brown in color (Dry Rot is also called brown rot). The decayed wood cracks into cubical pieces as the fungus destroys the cellulose in the cell walls. Although the brown lignin is largely unaltered, the structural strength is almost entirely lost (Singh and White 1997).

Dusting. The development of a powdered material at the surface of hardened concrete. Dusting can be caused by troweling condensation moisture or bleed water into the fresh concrete surface, resulting in a localized excessive water/cement ratio; inadequate curing, resulting in a soft surface skin; a weak surface layer due to excessive amounts of clay or silt in the concrete; inadequate ventilation during curing, allowing carbon dioxide to dissolve, turn to carbonic acid, and inhibit hydration; or rain, snow, or drying winds (American Concrete 1979; National 1978).

Eccentric Load. A leading cause of Overstress, the nonsymmetrical or off-center placement of loads on a structural member or system. The eccentricity can induce torsion or bending moments into members that were designed to resist only axial stress.

Edge Contact. A common cause of window glass breakage, caused by the glass lite coming into contact with the metal frame, which applies a concentrated or point load on the glass. Edge Contact can be an original installation defect but can also develop over time due to movement or racking of the frame, sliding on sloped glazing, or seismic activity.

Edge Defect. Nonuniform edges of tempered glass, such as chips, which can result in Spontaneous breakage if the defect impairs the tension zone of the glass. Edge Defects can result from manufacturing or handling. Conventional wisdom holds that Edge Defects occurring before tempering cause the glass to break during tempering or to develop a compressive zone around the defect, resulting in no consequence. But, in practice, as much as 20% loss of strength is possible.

Efflorescence. Salt accumulation (usually white) on the surface of a building material by drying or evaporation of the water from a salt-laden solution. Efflorescence most commonly consists of calcium and carbon sulfate and chlorides on the surface of masonry. The source is often unhydrated lime, which reacts with water and atmospheric carbon dioxide to form calcium carbonate and calcium sulfate. Efflorescence is generally not harmful but can be unsightly (American Concrete 1979; American Society 1990).

Electrochemical Corrosion. Oxidation of metal due to an electrochemical Reaction between the metal and oxygen in the environment, characterized by electrical current flow between anodic and cathodic areas in an electrolyte. The anodic Reaction is oxidation (Corrosion); the cathodic Reaction is reduction (protective). Electrochemical Corrosion is the most common cause of deterioration of unprotected iron and steel products. The corrosion product can vary, having a volume up to three times that of the base metal, and when restrained can exert over 4,000 psi of expansive stress (American Society 1990; Pludek 1977; Carper and Feld 1997).

Electrolytic Corrosion. See the preferred term Electrochemical Corrosion.

Embrittlement. To become more brittle—a process that results in a loss of ductility. Embrittled materials fail with less deformation than they would if not embrittled.

End Grain Attack. Deep intergranular Corrosion on a surface transverse to its rolling or extrusion direction. Rolling and extruding orient nonmetallic intrusions into stringers that are only exposed to Corrosion at the ends of the surrounding metal, resulting in deep pits into the cut ends.

Erosion. The gradual wearing away of the surfaces, edges, or corners of a material by the action of a flowing fluid, especially air and water (American Society 1990).

Erosion-Corrosion. Corrosion accelerated by the flow of a corrosive fluid, characterized by a grooved or wavy appearance, usually in a directional pattern (Von Fraunhofer 1974; Bosich 1970).

Etching. A form of Attack on glass from an alkaline solution. In flat glass used in glazing, the result is a permanent loss of visual clarity. In glass fibers used in reinforcing, the result is a loss of section (Necking) and therefore strength.

Euler Buckling. Flexural Buckling of a structural member (usually columns) with too high a ratio of length to radius of gyration. Euler Buckling results when the second-order moments (caused by the axial compressive stresses and the displacement) are equal everywhere to the internal bending resistance (Trahair 1993).

Exfoliation. A form of Delamination in natural stone resulting in Peeling, Scaling, or Flaking off of the surface in thin layers. Exfoliation is usually associated with water permeance, either resulting in Freeze-Thaw damage or the dissolution of the natural cement (American Society 1990).

Exfoliation Corrosion. See the preferred term Layer Corrosion.

Fatigue. The weakening (up to and including Fracture) of a material caused by repeated, cyclical, or alternating loads (stress reversals). Fatigue resistance is dependent on the level of stress and the toughness of the material. The characteristic appearance of a Fatigue failure plane is a relatively large smooth area and a smaller area with a roughened crystalline surface. The smooth region results from cyclical grinding of the fractured surface as the crack propagates. The crack propagates through the material until its cross-sectional area is insufficient to carry the tensile stress, causing a brittle failure (the roughened area). Fatigue Cracks initially propagate slowly and, if detected early, can be treated by taking appropriate remedial action. If the cracks are allowed to propagate unrestricted, they frequently initiate Brittle Fracture. In elastomeric sealants Fatigue is a significant cause of premature failure, leading to either Adhesion Failure or Cohesive Failure (Pludek 1977; American Society 1990).

Flaking. An early stage of Peeling, Exfoliation, Delamination, or Spalling (American Society 1990).

Flexural-Torsional Buckling. Buckling of unsymmetric and singly symmetric shapes (especially thin plates and columns) involving lateral deflection and twist, which occurs when the shear center and centroid do not coincide (Trahair 1993).

Fluid Migration. Whereas Bloom involves accumulation of fluid only on the surface of a material that is the source of the fluid, Fluid Migration describes those cases in which a fluid component moves to and accumulates onto or into an adjacent material. Fluid Migration can be observed as one of the following manifestations:

1. Darkening of the substrate, caused by the fluid being visible in or on the substrate. This is the classic Staining mechanism of porous stone substrates adjacent to certain sealants.
2. Waterproofing of the substrate (hydrophobic action), caused by the migrated fluid preventing water absorption. The propensity for this phenomenon is dependent on the porosity of the substrate. Typically, this condition is not permanent, because the low molecular weight fluids causing the effect will degrade and dissipate from Weathering.
3. Dirt Pickup on the substrate, when high molecular weight fluids migrate into or onto the substrates.

Flutter. Sustained coupled vibrations of a structure (especially bridge decks) at large amplitudes and at a common frequency due to crosswind excitation, involving instabilities in more than one degree of freedom (Irvine 1981).

Fogging. Loss of visual clarity in insulated glass units (IGUs) caused by moisture entry into the airspace. The airspace is intended to be sealed and desiccated, so Fogging is an indication of IGU seal failure. Repeated cycles of Condensation and evaporation of water inside the airspace cause salts (and sometimes Organic Growth) to deposit on the interior surface of the glass. Also, loss of visual clarity of laminated glass, due to a Reaction of the PVB interlayer and water, atmospheric pollutants, or sealants in contact with the edges of the glass lites. The characteristic pattern of laminated glass Fogging is toothed.

Fracture. To make or become discontinuous by means other than cutting, and the resulting

discontinuity. Fracture is usually associated with relatively brittle materials.

Fracture Cracking. Brittle Cracks that take place with little or no preceding plastic deformation. Low temperature (possibly as high as room temperature), stress and strain concentrations (including Micro-Cracking and Fatigue Cracks), and metallurgical composition are important factors influencing Fracture Cracking. These types of Cracks are often triggered by Impact or a sudden increase in load (American Society 1990).

Frame Shortening. The reduction in height of a building's structural frame due to elastic shortening and, for concrete frames, Creep and Shrinkage. For tall buildings, the accumulation of the floor-by-floor incremental shortenings can result in a significant total shortening, affecting floor levelness and causing distress in exterior cladding, elevator shafts, and other continuous systems. For each concrete column, the amount of shortening is dependent on the amount of reinforcement; the temperature and humidity during cure; the curing method and duration; the volume to surface ratio; and the time of application of loads (including the dead load of additional floors and live loads). Frame Shortening can be especially damaging when it occurs in conjunction with Moisture Expansion of masonry veneers (Russell and Corley 1977; Fintel et al. 1986).

Freeze-Thaw. Disintegration of a material (usually concrete or masonry) resulting from the cyclic actions of the freezing and thawing of absorbed water. In concrete aggregate, the properties that affect Freeze-Thaw resistance are porosity, absorption, type, and source (American Society 1990).

Fretting Corrosion. Corrosion that develops because of chafing of metal under load at contact areas between metals where there is vibration or slip of the metal surfaces. Fretting Corrosion is characterized by streaked longitudinal Staining. The necessary conditions for Fretting Corrosion to occur include: the metal surface is under load; there is a small degree of slip between the surfaces; and there is vibration at the interface. Two theories have been advanced that explain the Corrosion process: (1) Small particles of one or both surfaces are pulled off and then oxidized; and (2) The metal first oxidizes and then is pulled away from the metal (Bosich 1970; Pludek 1977; Mattsson 1989).

Frost Penetration. Freezing of water in concrete, preventing the water from hydrating the cement, thereby destroying the concrete's strength. Also, freezing of water in soil, resulting in heaving (Carper and Feld 1997).

Fungal Growth. A plant growth obtaining its nutrition by breakdown of organic matter, usually associated with the presence of dampness, for example, in timber. The plants are characterized by the absence of chlorophyll.

Galvanic Corrosion. Corrosion of metal exposed to a conductive solution that allows an electric (galvanic) current to flow between an anodic and a cathodic region. Three types of Galvanic Corrosion are recognized.

1. Bimetallic—the electrodes consist of dissimilar metals in contact. The less resistant metal becomes the anode and the more resistant metal the cathode. Tables of galvanic potentials have been published (see, for example, Bosich 1970).
2. Concentration Cells—the electrodes are of the same metal, but concentrations of electrolytes vary (see Concentration Cells Corrosion).
3. Thermo-Galvanic—the electrodes are of the same metal and the composition of the electrolyte is constant, but the temperatures differ at the electrodes (see Thermo-Galvanic Corrosion) (Bosich 1970; Mattsson 1989).

Graphitic Corrosion. A form of Selective Attack in which iron is leached out by Galvanic Corrosion, leaving a graphite network behind. Graphitic Corrosion, which is most often found in gray cast iron, is difficult to detect visually because the Corrosion product has the same volume and color of the original base metal. Therefore, it is usually found after a failure has occurred (Bosich 1970; Bradford 1993).

Hardening. An increase in the hardness of a material after Weathering or Aging, resulting in an adverse loss of ductility or elasticity. In elastomeric sealants, Hardening can lead to Adhesion Failure and occasionally Cohesion Failure.

Heave. Upward movement of soil (ground) or of a structure that it supports, caused by, among other factors, Frost Penetration, expansive clays, or hydrostatic pressure (such as water flow around sheet piles).

High-Temperature Corrosion. Corrosion of metal accelerated by ions breaking through the protective film on a passivated metal. The passification breaks down at high temperatures, and the Corrosion rate increases exponentially as the temperature increases (Pludek 1977).

Hinging. The failure (possibly leading to collapse) of a masonry arch with a low rise-to-span ratio caused by the opening of a space between two units (stones, bricks, etc.).

Honeycombing. A defective condition in concrete characterized by interconnected large voids resulting from loss or lack of paste. Honeycombing is usually caused by a lack of adequate vibration after placement, inappropriately low slump, or congestion of the reinforcing steel (American Society 1990).

Hydrogen Embrittlement, Hydrogen Attack, Hydrogen-Induced Delayed Brittle Fracture, Hydrogen Stress Cracking. Hydrogen penetration into metal can cause loss of ductility, loss of strength, cracking, or catastrophic Brittle Fracture below the yield strength of the material. The hydrogen atoms can diffuse into the metal before combining with other hydrogen atoms to form hydrogen molecules, which cannot readily diffuse. If the hydrogen atoms combine to form molecules in voids in the metal, very high internal pressures can develop, causing blisters and distortion. Hydrogen atoms can be introduced during fabrication, cleaning, pickling, phosphating, electroplating, autocatalytic processes, Corrosion, or as a cathodic protection reaction product. Metal that has been machined, ground, cold formed, or cold straightened subsequent to heat treatment is especially susceptible to Hydrogen Embrittlement. Hydrogen Embrittlement of steel prestressing tendons has been documented in bridge failures in the United Kingdom but does not seem to be prevalent in the United States (Raymond 1988; Von Fraunhofer 1974; Raymond 1988; Carper and Feld 1997).

Ice Crystal Formation. Freezing of confined water contained within voids of the pores of a porous material, such as mortar or concrete. The expansive force of the crystallization can destroy the material.

Impact. Nearly instantaneous load increase on a

structural system. Usually, Impact loading results from one body striking against another, but this term also applies to dynamic loads of bodies already in contact, such as the sudden tensioning of a wire rope.

Impingement Corrosion. Localized Corrosion resulting from Erosion by a liquid impinging on the metal surface, which wears away the protective film, making the impingement area anodic whereas the surrounding area is cathodic (Pludek 1977).

Inadequate Lip Pressure. In glazing, so low a force per unit length exerted by a glazing gasket against the glass or other substrate that water or air can penetrate. The industry generally recognizes a residual (after Weathering, Out-Gassing, Shrinkage, Hardening, and Embrittlement) lip pressure of 2.5 lb per linear inch as the minimum necessary to prevent leakage.

Intergranular Corrosion. Localized attack at grain boundaries (which are anodic relative to the grains) with little Corrosion of the grains. Intergranular Corrosion includes Weld Decay, Knife-Line Attack, and End Grain Attack. As the Corrosion proceeds, the grains fall out and the metal disintegrates.

Interstitial Condensation. An accumulation of moisture within porous masonry when moist air reaches its dew point. Interstitial Condensation is not itself a problem, but it is a source of liquid water within a material, which can lead to a number of failure mechanisms.

Knife-Line Attack. A form of Intergranular Corrosion in a narrow band in the base metal immediately adjacent to a weld, resulting from poorly controlled welding of stabilized steels (Bosich 1970).

Lamellar Tearing. A planar separation that develops within thick plates near certain large welds as high weld Shrinkage develops stresses across the plate thickness. Lamellar Tearing is associated with constraining details, in which the heat generated by welding cannot be dissipated; as the weld cools, microscopic cracking occurs. Impurities and nonhomogeneous steel cross-sections contribute to the problem (American Society 1990; Carper and Feld 1997).

Lateral-Torsional Buckling. Buckling of a structural member involving deflection and twist. When subjected to high bending moments (but below the yield strength), the compression zone deflects out of plane laterally, caus-

ing the member to twist. Lateral Torsional Buckling may control the capacity of beams with long unbraced lengths (American Institute 1986).

Layer Corrosion. A form of Intergranular Corrosion of metal characterized by swelling, resembling flaky pastry. Layer Corrosion can occur when the metal grains have been highly elongated in platelet shapes by cold work, and the grain boundaries are electrochemically different from the grains (Bradford 1993).

Leachate. The caustic salt deposited on a surface by Leaching. Leachate can interfere with adhesion of materials to be bonded to the surface or can cause Etching of glass.

Leaching. As a form of Corrosion, see the preferred term Selective Attack. Leaching also describes a process of water migration through alkaline materials transporting salts to the surface (Efflorescence) or onto adjacent surfaces such as glass. Also, Leaching describes the removal of sodium and other ions from a glass surface due to the presence of water, causing hazing (see also Moisture Attack).

Lippage. Out-of-plane nonmatching edges of adjacent stones, usually in floors, forming a lip along one or more edges. Lippage can be caused by original misinstallation or subsequent distortion of stones, such as warping from Thermal Hysteresis.

Local Buckling. Deflection of a member out of its original plane of approximately the same magnitude as the member's cross-section width. Local Buckling is controlled by width to thickness ratios of plate elements and usually occurs at the point of greatest in-plane compression, although multiple buckles may develop along members under essentially uniform stress (American Institute 1986; Trahair 1993).

Map Cracking. See the preferred term Pattern Cracking.

Masonry Corrosion. An imprecise term for the gradual wearing away of the exposed surface of masonry by external actions of chemicals or other forms of physical Deterioration. See the preferred terms Attack, Deterioration, and Erosion (American Society 1990).

Microbial Corrosion. Chemical attack of metals (primarily underground) from by-products (sulfuric acid, carbonic acid, hydrogen sulfide, ammonia) of bacteria, fungi, and molds. Also, microbes can directly attack organic coatings, depassify metal surfaces, and induce

Corrosion cells. Microbes set up an oxygen Concentration Cell; under the biofilm, the oxygen concentration is very low, creating an anodic region that is subject to Corrosion (Pludek 1977; Scott and Davies 1992).

Microcracking. Small internal cracks in a material (especially concrete), which may form a considerable network before becoming visible at external surfaces, leading to Disintegration. Microcracking in concrete is categorized into two types: (1) At the aggregate-mortar interface, caused by shear and tensile stresses at the interface due to early volumetric change (hydration and drying Shrinkage). This mechanism is not load dependent; and (2) Cracks within the paste, which are generally caused by external loading (Nawy 1996; Mindess and Young 1981).

Mildew. A type of Mold that occurs on wood, concrete, or other absorbent materials in a warm, moist, and poorly ventilated environment.

Missile Impact. The primary cause of window glass breakage—impact by wind-borne missiles. The missiles usually consist of roof gravel or other small, wind-borne debris.

Moisture Attack. Staining, Etching, or Deterioration of glass surfaces caused by long-term exposure to water, such as from irrigation sprinklers. Because glass ingredients are alkaline, ion exchange with a thin layer of water will cause the water to become alkaline, which can Attack the glass, leaving a smoky haze on the surface. Moisture Attack can be exacerbated by air pollution.

Moisture Expansion. Volume expansion of construction materials resulting from the absorption or adsorption of water. Restrained Moisture Expansion can cause high stresses (sufficient to cause cracking) or displacement (such as arching of masonry floors or walls). Some Moisture Expansion is reversible (such as in wood and a portion of the Moisture Expansion in brick masonry), whereby the material will shrink and swell depending on the moisture content in its pore spaces. In desiccated materials, such as fired-clay bricks, some of the Moisture Expansion is irreversible without heating to drive off the absorbed water. Average Moisture Expansion of clay bricks is 0.02%, but depending on the particular clay used and its manufacturing history, the Moisture Expansion can vary from

near 0 to over 0.05% (Carper and Feld 1997; Grimm 1975).

Mold. One of several woolly or powdery fungal growths that form on the surface of materials in damp, stagnant atmospheres, causing patchy discoloration that is usually green, gray, or black but occasionally pink, red, or yellow. On wood, Mold usually has a negligible effect on strength but may cause some loss in toughness and an increase in porosity. Mold may discolor and seriously weaken paper and fiber-based products (Singh and White 1997).

Mudcracking. A rectangular pattern of Cracking in paint films caused by Shrinkage during drying.

Necking, Neck Down. A reduction in section caused by any of several processes, including Attack, Corrosion, or Etching.

Nickel Sulfide Inclusion. In tempered glass (and more rarely in heat-strengthened glass), an impurity that can rupture the lite upon thermal expansion. Liquid droplets of nickel sulfide (NiS) in molten glass solidify as the glass cools, resulting in small, brassy polycrystalline spheres, typically 0.1–1 mm in diameter. During glass tempering, the NiS is transformed and "trapped" in the high-temperature alpha phase. With the passage of time (hastened with elevated temperature), its chemical phase changes; it slowly inverts to the low temperature beta phase, accompanied by a volumetric expansion of 2–4%. The crystals are anisotropic, so the expansion varies along each axis. The expansion causes internal Stress Concentrations that can spontaneously Rupture the glass lite if the inclusions are located in the interior (tension) zone of the glass (see Spontaneous Breakage). Conventional wisdom holds that only fully tempered glass is subject to Spontaneous Breakage from NiS Inclusions, but there are many documented cases of breakage from NiS Inclusions in heat-strengthened glass, as well. Although NiS Inclusions sometimes can be located and their size measured in situ, there is no known way to determine the phase of the inclusions. The characteristic appearance of an NiS-induced breakage pattern, if recovered intact, is a "butterfly wing": two large symmetrically opposed polygonal crystals around a smaller fragment with a shatter plane near its mid-depth (Brungs and Sugeng 1995).

Nitrate Attack. Attack of concrete by concentrated nitric acid, potassium nitrate, or ammonia nitrate, causing Deterioration.

Off-Gassing. See the preferred term Out-Gassing.

Oil-Canning. Buckling of sheet metal or thin materials, usually caused by thermal stress and resulting in deformations. Such deformations can be so small as to be impossible to measure outside of a laboratory and yet cause significant aesthetic detraction.

Opacifier Failure. Distress in the opacifier coating on spandrel or other nonvision glass, resulting in Color Change, Peeling, or other dysfunction. Because there are many types of opacifiers, there are many possible causes of failure. Most opacifier problems have been associated with Shrinkage, Weathering, oxidation, or Adhesion Failure of coatings and adhered films that are moisture-, heat-, or light-sensitive.

Organic Growth. Algae, mildew, and other micro-organism growth on a surface, resulting in aesthetic distress or Decomposition of the substrate. This growth may resemble Dirt Pickup.

Out-Gassing. The evaporation of volatile compounds from a chemical-based product, such as rubber or plastic. In addition to health concerns from volatile organic compounds, Out-Gassing can cause physical performance problems including Shrinkage or Hardening of the product and can result from Weathering.

Overstress. A loading condition on a structural member or system that results in any part being subjected to an actual stress level above one of its maximum design service stress levels, or in the case of an understrength component, above its actual capacity.

Parting. See the preferred term Selective Attack.

Pattern Cracking. Crazing in a concrete surface in a distinct pattern resulting from a decrease in volume of the material near the surface or an increase in volume of the material below the surface, or both. Factors that contribute to propagation of Pattern Cracking include, heat, Freeze-Thaw, chemical reactivity of aggregate, and aggregate-paste bond (which is dependent on grading, particle shape and surface texture, and chemical or residual oxide composition) (Marek 1991; Mindess and Young 1981).

Pattern Loading. An arrangement of loads on a structural system that results in additive stresses, such as gravity loads on alternating bays of a continuous beam. Pattern Loading can cause Overstress, even if the total load on each section is within the capacity of the individual section.

Peeling. The loss of adhesion and Deterioration of a coating, paint, or glaze from the substrate.

Pillowing. Warping of thin cladding panels, giving them an inflated appearance. Theoretical causes of Pillowing include differential Thermal Expansion and differential Moisture Expansion between the inner and outer faces of thin panels. The resulting differential strains cause bowing. Metal panels can exhibit Pillowing if the thickness is insufficient, if the span is too large, or from residual stress induced by fabrication rollers. Thin ceramic spandrel panels can also exhibit Pillowing.

Pinpoint Corrosion. Localized Corrosion development on ferrous metals coated with either organic or inorganic coatings, as a result of normal Weathering of the coating (Tam and Stiemer 1996).

Pitting. On glass surfaces, small voids typically caused by welding sparks. Sparks landing on glass cause localized Thermal Shock which, in turn, causes Pitting.

Pitting Corrosion. Extreme localized attack caused by anodic areas remaining small, discrete, and isolated, whereas the surrounding surface remains cathodic. Pitting Corrosion begins as a localized breakdown of the protective film on the metal caused by a defect, impurity, scratch, or weak spot. The pits remain active if their pH is more acidic than the surrounding metal. Pitting Corrosion can also be subsurface, resulting in larger voids in the metal (Bradford 1993).

Plastic Hinge. A yielded zone that forms in a structural member when the resisting moment of a fully yielded cross section is attained. The member is assumed to rotate as if hinged, except that it is restrained by the plastic moment (LRFD 1986).

Point Contact. In stone veneer connections, point loading of an anchor on the edge of a kerf due to improper design or construction (Hooker 1995).

Ponding. The accumulation of water in a depressed area, usually on a low-slope roof or deck. Ponding can create relatively harmless "birdbaths," or it can be a collapse mechanism. As the weight of additional ponded water causes downward deflection of structural members, additional water can accumulate in the deeper depression. Eventually, the incremental deflections either converge (reach equilibrium) or diverge (collapse).

Pop-Out. See the preferred term Spall.

Pull-Out. Failure of an embedded fastener (such as a nail or expansion anchor) by pulling out of its hole due to a lack of friction resistance. In wood, Pull-Out may be caused by Connection Loosening. In masonry, a Crack may widen the fastener hole causing loosening (Michaud 1996).

Pull-Through. Failure of a torque-controlled expansion anchor by pulling the conical expansion portion through the sleeve (Michaud 1996).

Punching Shear. The most common mode of failure for flat-plate concrete structures, whereby a column or concentrated load punches through the slab. Punching Shear failures, which usually occur suddenly and with relatively low applied loads, are governed by the column plan dimensions, the strength of the concrete, and the effective depth of the slab. For symmetric loading in the interior of a slab, Punching Shear is characterized by initial flexural cracks around the column perimeter. At higher loads, flexural cracks form radially from the column. A series of circumferential cracks form around the column at 60–80% of ultimate load. At this stage, the shear forces are transferred through dowel action, aggregate interlock, and the compression zone of the slab, which is in a triaxial state of stress. Finally, a faulted crack develops around the column, with a typical shear crack angle of 30° to the plane of the slab, displacing a truncated cone-shaped element (Carper and Feld 1997; Chana 1991).

Reaction. An unintended chemical reaction between two materials in contact or between a material and exposure to environmental chemicals. Reactions often lead to Deterioration or Adhesion Failure but sometimes to other manifestations such as exudate or Color Change of the materials.

Relaxation. The intrinsic, time-dependent reduction (from 3 to 12%) of the initial stress applied to the prestressing tendons in prestressed concrete. Relaxation is caused by the realignment of the steel fibers under stress. Other causes of loss of prestress, which should be differentiated from Relaxation, include Shrinkage, Creep, and elastic shorten-

ing. The amount of Relaxation is a function of time, ambient temperature, class of steel, and the ratio of initial stress to yield stress. Relaxation is similar to Creep except that Relaxation refers to a change in stress under constant strain, whereas Creep refers to a change in strain under constant stress (Lin and Burns 1976; O'Brien 1995).

Reversion. The loss of the elastomeric property of a formerly cured organic sealant, which sometimes appears melted or gummy. After reverting, the sealant is no longer a cured rubber; it can be reshaped. Not all scientists agree on whether the polymer is truly reverting to an earlier, precured state or advancing beyond cure to a new state. Probably, Reversion is not just one mechanism. Depending on the specific product formulation, Reversion appears to include processes caused by each of the following alone or acting in combination: heat, ultraviolet radiation, and moisture (either as a liquid or vapor).

Ringworm Corrosion. Corrosion in a band or ring around a pipe near the threaded or flanged ends caused by reheating to thicken or thread them. Near the pipe ends, the metal is only heated enough to transform the metallurgy to spheroidite (not the typical tempered martensite or bainite, which are reasonably Corrosion resistant). Spheroidite has a ferrous matrix with large spheres of iron carbide (the cathodes), which corrode (Bradford 1993).

Rising Damp. See the preferred term Capillary Rise.

Roll-Out. In glazing, the loss of glass lites from window frames due to deformation of perimeter gaskets under negative wind loading.

Rot. See Dry Rot, Wet Rot, Soft Rot, and Decay.

Rundown. The accumulation of a fluid component of a material on vertical or sloping surfaces. This aesthetic failure mechanism is often confused with dirt run-off patterns from rainfall, such as at window corners, which are more common.

Rupture. That stage in the development of a Fracture at which instability occurs and incremental additional propagation diverges.

Salt Fretting. A pattern of Erosion of stone caused by dissolved salts migrating toward the stone surface, crystallizing behind an indurated crust. Crystallization can only occur with saturated solutions.

Sand Pocket. A localized defect in a concrete

member containing fine aggregate without cement (American Concrete 1979).

Scaling. Localized Flaking or peeling away of the surface of concrete or mortar, which is classified by severity from light to very severe, depending on the degree of loss or exposure of coarse aggregate (American Concrete 1979; American Society 1990).

Scour. Erosion of a structural member, such as bridge piers and abutments, caused by running water (usually of high velocity), which mobilizes the stream bed material (Murillo 1987).

Scumming. See the preferred term Moisture Attack.

Segregation. The differential concentration of the components of mixed concrete, resulting in nonuniform proportions of the mass. Usually, Segregation refers to a horizontal Stratification caused by gravity (see also Separation, Stratification) (American Concrete 1979; American Concrete 1990).

Seizing. The development of extreme friction in a pin connection or hinge due to a lack of lubrication or Corrosion. As the connection restricts free rotation, unintended stress is transferred through it, such as bending moments induced in axial members, which can result in Overstress.

Selective Attack. The selective removal of one element, usually the more noble, from an alloy by Corrosion. This type of Corrosion leaves the alloy weakened and susceptible to Stress Corrosion Cracking and Corrosion Fatigue. Selective Attack is a slow process and only occurs in certain alloys under specific conditions (Craig 1989; Bradford 1993).

Separation. A synonym for Segregation. More specifically, Separation may also indicate the tendency, as concrete passes from the unconfined ends of a chute or similar apparatus, for coarse aggregate to separate from the concrete and accumulate to one side (American Concrete 1990).

Settlement. A downward movement of the soil or of a structure that it supports. Initial (early) Settlement is caused by deformation of soil particles under added weight or collapse of subterranean structures. Over time Decay of organic matter in the soil, Consolidation, and soil Shrinkage due to dewatering can cause additional Settlement (Sowers and Sowers 1970).

Shattering. The Disintegration of unreinforced PVC single-ply roofing membranes, resulting in spontaneous Crack development, usually in cold weather. Shattering is associated with plasticizer Out-Gassing and exudation, which causes Shrinkage and causes the membrane to Embrittle. Roof ballast can exacerbate the Embrittlement by absorbing plasticizer from the membrane. Empirical data indicate that Shattering occurs after 4–13 yr of exposure, with a median of 8.5 yr. The Shrinkage increases strain, whereas the Embrittlement increases the membrane's coefficient of Thermal Expansion (Cash 1992; National 1990).

Shrinkage. A decrease in length, area, or volume that occurs due to an internal process, as opposed to the response of a material to an external action. In concrete, early-age Shrinkage is attributed to the evaporation of excess water (drying Shrinkage), autogeneous Shrinkage due to the hydration of the cement, and the temperature gradient caused by the internal heat of hydration. Later Shrinkage is caused by continuing hydration and Carbonation (Rollings 1993).

Side Sway. A form of racking of cantilevered columns with no cross bracing, usually during erection. Side Sway is of particular concern in lift-slab construction.

Sinkhole. A depression in soil usually associated with acidic groundwater dissolving limestone strata, resulting in gradual Subsidence or sudden collapse of the overlying material (Sputo 1993).

Slab Curling. A distortion (typically warping upward) of a concrete slab on grade caused by differential Shrinkage between the top and the bottom of the slab, friction between the slab bottom and underlying material, and thermal gradients (Rollings 1993).

Sliding. The failure (possibly leading to collapse) of a masonry arch caused by a unit (brick, stone, etc.) slipping out of place.

Slip-Stick Movement, Stick-Slip Movement. Nonuniform thermal expansion or contraction of a system characterized by the sudden movement ("slip") relieving accumulated ("stuck") stress. Slip-Stick Movement is especially common in aluminum curtain wall framing. The sudden release of energy can dislodge intersecting members and is often audible to the point of causing alarm. The restriction to sliding can be caused by alkaline run-off water coming into contact with aluminum members, such as from concrete placements above installed curtain wall components.

Slope Stability Failure. The downward and outward movement of soil under the action of gravity due to shear stress—a landslide. The stability of a slope is a function of the type and strength of soil layers, the ground-water table, and the geometry of the slope. A slope can lose stability and collapse due to an increase in shear stress or a decrease in shear strength. Specific causes include the slope's own weight (especially from Creep or vegetation); the influx of water reducing the shear strength of the soil; dynamic loading (such as seismic); Erosion; lateral earth pressure; undercutting; or a change in geometry. Failures are classified into six main categories: falls, topples, slides, lateral spreads, flows, and complex (Fang 1991; Carper and Feld 1997; Terzaghi 1950; Varnes 1978).

Soft Rot. Decay of wood caused by a fungus that, although common in waterlogged hardwoods, is less damaging and less detectable than Dry Rot or Wet Rot. When dry the surface appears as though it has been lightly charred, with profuse fine cracking and fissuring both with and across the grain (Singh and White 1997).

Spall. A fragment loosened (or completely dislodged) from the surface of masonry (usually concrete), which leaves a shallow, typically conical depression. Also, the action of causing a Spall to form. Spalls may be caused by Impact, Corrosion of embedded steel, Cement-Aggregate Reactions, internal ice crystal formation, or Cryptoflorescence (American Concrete 1979; American Society 1990).

Spandrel Volatilization. In shadow-box glazed spandrels, the condensation of volatile organic compounds (VOCs) on the inside face of the glass. The intense heat within glazed spandrels can cause VOCs to vaporize, which later condense as the air cools. Sources of VOCs in shadow boxes include tapes; adhesives; contaminants from construction, such as machine oil; and plasticizer Out-Gassing from plastic components. The condensate may chemically attack the opacifying film or coating, resulting in Staining. Even if there is no reaction, a thin film of condensate (between 110 and 220 nm thick) can cause vivid rainbow-colored iridescence from the light transmission interference. The Discoloration often appears as a bright band around the pe-

rimeter of the spandrel, which tends to cool first from metal-framing conduction.

Split. A break in a material, approximately parallel with the natural grain or cleavage of the material. Splitting in wood can be caused by drying Shrinkage, increasing the stress on connections.

Spontaneous Breakage. In glass the sudden Rupture and Disintegration of heat-strengthened or fully tempered lites, especially due to an undetermined cause. Typical causes of Spontaneous Breakage include Impact (including Missile Impact); Edge Contact, Thermal Shock; or latent Rupture from manufacturing defects such as impurities (including Nickel Sulfide Inclusions), bubbles, scratches, Cracks, and Edge Defects. Scratches, Cracks, and Edge Defects can also occur during transportation, handling, installation, and service.

Stain. The physical manifestation resulting from Staining. Also, the action of causing a Stain.

Staining. A change in the appearance of a material caused by the accumulation of a foreign material on its surface or embedded within it (see also Discoloration).

Stratification. The Separation of overwet or overvibrated concrete into horizontal layers with increasingly lighter material toward the top. Water, laitance, mortar, and coarse aggregate will tend to occupy successively lower positions in that order. Also, a layered member resulting from placing successive batches that differ in appearance (American Concrete 1990). Also, the formation of hot, stagnant air under the apex of sloped glazing or skylights, which can cause glass breakage due to Thermal Expansion.

Stray Current Corrosion. Electrochemical Corrosion from uncontrolled electrical currents from extraneous sources through unintended paths, such as under water or through soil.

Stress Concentration. A stress at a particular place in a component that is significantly higher than the average due to the geometry of the component, such as corners, offsets, or dimensional changes. Failures are often initiated near Stress Concentrations when the localized stress of one type is higher than the component's capacity to resist that type of stress, even though the average stress in the member is below its overall capacity.

Stress Corrosion Cracking. Crack development in metal caused by the combined effects of localized Corrosion (typically Pitting or Crevice

Corrosion) and sustained tensile or residual stress (such as that caused by cold working or welding). As the Corrosion progresses and the metal loses section, the stress becomes more concentrated, accelerating the cracking (Pludek 1977; Bosich 1970).

Subsidence. The downward displacement of soil (overburden) lying above an underground excavation or weakened stratum, or adjoining a surface excavation. Subsidence usually refers to vertical displacement but can also refer to lateral displacement generated by adjacent downward displacement. Principal factors contributing to natural subsidence include soil compaction, soil Shrinkage, lowering of water table, development of subterranean voids, and tectonic and volcanic activity. Sinkholes are the most common form of natural subsidence in limestone regions (Whittaker and Reddish 1989).

Sugaring. An imprecise term describing the coarse grain structure of natural stone or the loosening of the grain boundaries over time such that upon fracturing the stone resembles sugar. Also, a characteristic of some masonry indicative of gradual surface Disintegration, sometimes in a powdery condition (American Society 1990).

Sulfate Attack. In concrete, chemical Reaction between sulfate solutions (resulting from, e.g., alkali magnesium and chloride sulfates in groundwater) and hardened cement paste. The sulfate reacts with the calcium hydroxide in the cement paste to form gypsum, and the gypsum reacts with the hydrated calcium aluminate to form ettringite (also sometimes called "cement bacillus"). The products of the reaction (crystal growth) can lead to internal expansive forces and disintegration of the concrete. Sulfate attack resulting in damage due to salt crystallization is commonly observed in areas where water migrates through concrete and then evaporates. In general, sulfate resistance is dependent on the type of cement used, the quality of the concrete, and the surface protection of the concrete (American Society 1990; Day 1995).

Swelling. The volume expansion of soil (especially plastic clays) due to the absorption of water. For materials other than soil, see the preferred term Moisture Expansion.

Telegraphing. A pattern that forms on the surface of a wall or slab mimicking a concealed pattern on the interior of the wall or slab,

such as the location of studs or reinforcement. Usually, telescoping is caused by vapor migration or differential thermal conductivities.

Thermal Contraction. A decrease in volume of a component as a response to a decrease in temperature (see also Thermal Expansion).

Thermal Cracking. In concrete, cracking resulting from differential thermal strains due to external heating or from internal heat of hydration (Springenschmid 1994).

Thermal Expansion. An increase in volume of a component as a response to an increase in temperature. The idealized behavior of a material is that its Thermal Expansion is linearly proportional to the change in temperature and its coefficient of thermal expansion—a material property. The degree of restraint of the component, however, will affect the actual Thermal Expansion, and the coefficient can be affected by moisture absorption. Furthermore, Thermal Expansion is a nonlinear function of temperature for some materials (Widhalm et al. 1996; Widhalm et al. 1997).

Thermal Hysteresis. Irreversible loss of strength, usually manifested as visible deformation of thin (less than 2 in.) marble cladding panels due to anisotropic Thermal Expansion. Usually, the panels become convex outward but can become concave inward for certain boundary and gradient conditions. Upon heating, the marble crystals expand nonlinearly and anisotropically, and upon subsequent cooling do not return to their original length. The degree of hysteresis along each axis differs and may result from the preferred orientation of the grains due to bedding planes or metamorphosis. The compressive region of the warped panel experiences Creep, which reinforces the warped shape. In addition to Warpage, Thermal Hysteresis is associated with a gradual loss of strength as the grain boundaries loosen (Sugaring) because of the anisotropic thermal expansion of the calcite grains. Several theories have been advanced to explain the deformations during heating, including (1) loss of residual moisture; (2) relief of residual stress; and (3) loosening of marble grain boundaries. Laboratory testing indicates that the majority of thermally induced disaggregation occurs during the first heating cycle; subsequent temperature cycles do not lead to any further considerable damage. Therefore, thermal cycling cannot solely explain the deformations observed in the field. There remains some debate regarding the importance of moisture in initiating Thermal Hysteresis, but there is no doubt that as it continues the porosity of marble increases, thus allowing secondary failure mechanisms involving moisture (such as Freeze-Thaw and Chemical Weathering) to exacerbate the attendant loss of strength. Glass fiber-reinforced concrete (GFRC) panels behave similarly, perhaps for analogous reasons (Widhalm et al. 1996; Widhalm et al. 1997; Bortz et al. 1988; Cohen and Monteiro 1991).

Thermal Shock. The force, arising out of thermal expansion or contraction, that causes disruption of a material on sudden heating or cooling. Glass is vulnerable to breakage from Thermal Shock due to uneven heating and shading patterns and to Pitting from sparks. Generally, maximum thermal stresses occur when less than 25% of the glass area and more than 25% of the perimeter length is shaded. Horizontal, vertical, and diagonal shading patterns are not as harmful as combinations of these.

Thermo-Galvanic Corrosion. Galvanic Corrosion caused by a thermal gradient in the metal, which reduces the metal's resistance to Corrosion. Uneven heating causes a metal to become polarized, forming an anode and a cathode. The protective film in a passivated metal breaks down at higher temperatures; the Corrosion rate increases exponentially as the temperature increases (Pludek 1977).

Torsional Buckling. Buckling of a structural member (usually symmetric shapes with thin plate elements) involving twist deformation. Torsional Buckling also refers to the special type of stability failure of towers and tanks, which tend to fail macroscopically by twisting the entire structure, with the supported mass screwing downward. For this type of stability failure, no individual member may buckle. For an individual member, Torsional Buckling results when second-order moments (caused by axial compressive stress and twist) are equal everywhere to the sum of the internal torsion resistances. A member is susceptible to Torsional Buckling if its shear center and centroid coincide. Especially vulnerable are cantilever and continuous beams, which have a compression zone (bottom) unbraced by the supported structure above (Trahair 1993; Allen and Bulson 1980; American Institute 1986).

Undermining. Erosion of soil from beneath a structure or foundation, which can result in collapse.

Uniform Attack. Corrosion characterized by a chemical or electrochemical Reaction that proceeds uniformly over the entire exposed surface. Uniform Attack can result in the formation of a protective layer that inhibits further Corrosion but often results in a loss of section as the base metal is corroded. The extent of Uniform Attack is measured as weight loss per unit area for the average depth of penetration. Uniform Attack is often associated with contact with an acid, high temperature oxidation in a relatively dry atmosphere, or sulfidation (Craig 1989; Bosich 1970; Pludek 1977; Mattsson 1989).

Vein Deterioration. In marble the Deterioration of the natural veins by chemical Attack (usually acids).

Vortex Shedding. The repetitive creation of circular air currents (vortices) on the leeward side of a structure, especially at the corners and edges. The vortices, which create high dynamic positive and negative pressures, tend to spin away (shed) from the structure after creation.

Walking. Any of a variety of phenomena involving incremental one-way movement. Examples include (1) pavement slabs Walking away from the crown of a roadway by gravity; (2) bearing pads Walking off their seats under cyclic loading; and (3) fasteners Walking out of threaded holes (unthreading) due to thermal expansion and contraction (see also Thermal Hysteresis).

Warp; Warping. Curvilinear deformation of a building component. Warping can be caused by, among other things, differential Moisture Expansion and Thermal Hysteresis.

Water Absorption. The unintended absorption of water into an elastomeric sealant. The absorbed water can then diffuse into the joint substrates (especially stone), creating a "wet joint" appearance (an aesthetic failure) and the potential for Freeze-Thaw damage of the substrates.

Water Pocket. In concrete, voids (initially filled with water) along the underside of aggregate particles or reinforcing steel, which formed during Bleeding.

Wax Bleed. In hardboard (such as that used for simulated wood siding), the migration of wax (used as a waterproofing medium for the cellulose fibers) to the surface of the boards. Wax

Bleed appears as oily "clouds" on the surface, especially on painted boards where the pigment absorbs the wax.

Weather Checking. See the preferred term Crazing.

Weathering. Degradation due to exposure to the weather. Weathering factors include the following: ultraviolet radiation; temperature; moisture (solid, liquid, and vapor); wind; ozone; carbon dioxide; pollution; and Freeze-Thaw.

Web Buckling. Buckling of a plate girder caused by bending in the plane of the web, which reduces the efficiency of the web to carry the bending moment. Web Buckling should be differentiated from Web Crippling, which occurs at intersecting members (American Institute 1986).

Web Crippling. Plastic deformation of the web of a structural member at an intersecting member caused by an excessive stress concentration (American Institute 1986; Salmon and Johnson 1990).

Web Shear Buckling. Buckling of the web of a structural member along the plane of compressive stress due to high shear stress, usu-

ally in short-span beams (American Institute 1986).

Weld Decay. Intergranular Corrosion in a band in the base metal (usually stainless steel), somewhat removed from a weld, in regions adjacent to grain boundaries, caused by poorly controlled heat treatment. The heat from the welding process causes the carbon to form chromium carbide, which is insoluble in the parent metal and is precipitated from solution. The chromium-depleted regions cause a large potential difference between the anode and the cathode. The chromium carbide region is resistant to Corrosion. Weld decay does not occur in low carbon (less than 0.02%) steels (Von Fraunhofer 1974).

Wet Rot. Decay of wood caused by any of a variety of fungal species. Wet Rot (also called white rot) destroys both the cellulose and the lignin, leaving the color of the wood largely unaltered but producing a soft or spongy texture. Wet rot occurs in persistently damp conditions and needs an optimum moisture content of 50–60% (Singh and White 1997).

Yield. Onset of nonelastic behavior in a structural member caused by an actual stress level above the material's elastic limit.

TABULAR INDEX OF FAILURE MECHANISMS

Element	Manifestations of Distress							Materials and Systems						Reference
	Aesthetic	Strength/stress/break/rupture	Movement/distortion/strain/stability	Change in properties	Change in volume	Loss of function	Deterioration	Structural systems, foundations, and soil	Concrete, masonry, and stone	Metals	Wood and wood products	Plastic, coatings, sealants, and rubber	Glass and glazing	Case Histories (page no.)
Abrasion	X	X			X		X		X	X	X	X	X	
Adhesion failure		X				X	X					X		32, 62
Aging	X	X		X			X		X			X		20, 22
Algae	X						X		X		X	X		58
Alkali-carbonate reaction		X		X	X		X		X					
Alkali-silica reaction		X		X	X		X		X					
Alligatoring	X	X			X		X					X		
Attack	X				X		X		X	X	X	X	X	
Bitumen floating						X	X					X		
Bleeding	X						X		X			X		
Blister	X					X						X		
Bloom	X											X		
Bond failure		X	X						X					
Breadloafing	X					X	X					X		
Brick suction		X			X				X					28
Brittle fracture		X							X	X				
Bubbling	X			X		X						X		
Buckling			X					X	X	X	X			30
Capillarity						X	X		X		X			
Capillary rise	X						X		X		X			
Carbonation				X	X				X					
Cavitation corrosion		X			X		X			X				
Cement-aggregate reaction		X		X	X		X		X					
Chalking	X						X					X		
Checking	X	X		X	X						X			
Chloride attack				X	X		X		X					
Chloride-accelerated corrosion		X			X		X		X	X				
Cissing	X				X	X	X					X		
Cohesion failure	X					X						X		
Cold joint		X							X					
Color change	X								X	X		X	X	
Concentration cells corrosion		X			X		X			X				
Condensate corrosion		X			X		X			X				
Condensation						X	X		X	X	X	X	X	

95

Element	Manifestations of Distress							Materials and Systems						Reference
	Aesthetic	Strength/stress/break/rupture	Movement/distortion/strain/stability	Change in properties	Change in volume	Loss of function	Deterioration	Structural systems, foundations, and soil	Concrete, masonry, and stone	Metals	Wood and wood products	Plastic, coatings, sealants, and rubber	Glass and glazing	Case Histories (page no.)
Cone failure		X							X					
Connection loosening		X	X		X					X				
Consolidation			X		X			X						
Consumption							X		X		X	X		
Contaminated aggregate		X							X					
Contaminated mixing water		X							X					
Corrosion	X	X		X	X	X	X			X				26, 58, 68
Corrosion fatigue		X		X	X	X	X			X				
Corrosion undercutting		X		X	X	X	X			X		X		
Cracking	X	X				X			X	X				12, 26, 28, 44, 54, 56, 58, 66, 74
Crazing	X				X		X		X			X		
Creep			X		X				X		X			54, 56
Crevice corrosion		X			X	X	X			X				26
Cryptoflorescence		X					X		X					
Crystal packing		X		X					X					22
D-cracking	X				X		X		X					
Dealloying		X		X			X			X				
Decay	X	X		X			X				X	X		
Decomposition	X	X		X			X				X	X		
Dedolomitization		X		X			X		X					
Degradation	X			X			X		X	X	X	X	X	
Delamination		X										X		
Delayed ettringite formation		X		X					X					
Deposit corrosion		X		X	X	X	X			X				
Deterioration	X	X		X		X	X		X			X		54, 66
Diagonal cracking		X							X					16
Diagonal tension failure		X							X					16
Differential movement			X		X			X	X	X	X	X	X	54, 56
Differential settlement			X					X						30
Differential support		X						X	X	X	X			8
Dirt pickup	X											X		
Discoloration	X								X	X	X	X	X	
Discontinuity		X								X				
Disintegration		X		X			X		X	X	X	X		
Downdrag		X						X						
Dragdown		X						X						

Element	Aesthetic	Strength/stress/break/rupture	Movement/distortion/strain/stability	Change in properties	Change in volume	Loss of function	Deterioration	Structural systems, foundations, and soil	Concrete, masonry, and stone	Metals	Wood and wood products	Plastic, coatings, sealants, and rubber	Glass and glazing	Case Histories (page no.)
Drift			X					X						
Dry rot	X	X		X		X	X				X			
Dusting	X								X					
Eccentric load		X						X	X	X	X			74
Edge contact		X											X	38, 70
Edge defect		X											X	38
Efflorescence	X								X					54
Electrochemical corrosion	X	X		X	X	X	X			X				
Embrittlement		X		X					X	X		X		22, 24, 50, 52
End grain attack		X		X	X	X	X			X				
Erosion	X	X			X		X	X	X	X		X		
Erosion-corrosion		X		X	X	X	X			X				
Etching		X			X	X			X				X	
Euler buckling			X					X	X	X	X			
Exfoliation	X				X		X		X					
Fatigue		X							X	X	X	X		14
Flaking	X				X		X		X			X		
Flexural-torsional buckling			X					X	X	X	X			
Fluid migration	X								X			X		46
Flutter			X					X						
Fogging	X												X	70
Fracture		X							X	X	X		X	14, 20, 70
Fracture cracking		X						X	X	X				
Frame shortening			X		X			X	X					56
Freeze-thaw							X		X					46, 58
Fretting corrosion	X	X		X	X	X	X			X				
Frost penetration		X	X		X			X	X					
Fungal growth	X								X		X	X	X	
Galvanic corrosion	X	X		X	X	X	X			X				
Graphitic corrosion		X				X	X			X				
Hardening				X			X					X		64
Heave			X					X						
High-temperature corrosion		X		X	X	X	X			X				
Hinging		X							X					
Honeycombing	X	X							X					58
Hydrogen embrittlement		X		X						X				
Ice crystal formation		X					X		X					26, 54, 68

97

Element	Manifestations of Distress							Materials and Systems						Reference
	Aesthetic	Strength/stress/break/rupture	Movement/distortion/strain/stability	Change in properties	Change in volume	Loss of function	Deterioration	Structural systems, foundations, and soil	Concrete, masonry, and stone	Metals	Wood and wood products	Plastic, coatings, sealants, and rubber	Glass and glazing	Case Histories (page no.)
Impact		X						X						
Impingement corrosion	X	X		X	X	X	X			X				
Inadequate lip pressure						X	X					X	X	50, 64
Intergranular corrosion		X		X	X		X			X				
Interstitial condensation					X				X					
Knife-line attack		X		X	X	X	X			X				
Lamellar tearing		X								X				
Lateral-torsional buckling			X					X	X	X	X			
Layer corrosion	X	X		X	X	X	X			X				
Leaching, leachate	X			X					X				X	
Local buckling			X					X		X				
Microbial corrosion		X		X	X	X	X			X				
Microcracking		X					X		X					
Mildew	X						X		X		X	X		
Missile impact		X											X	38
Moisture attack	X												X	
Moisture expansion		X			X				X		X			54, 56, 68
Mold	X					X			X		X	X		
Mudcracking	X	X			X	X	X					X		
Necking, neck down		X			X						X		X	22, 26
Nickel sulphide inclusion		X											X	38
Nitrate attack					X		X		X					
Oil-canning	X		X							X				
Opacifier failure	X			X		X	X						X	
Organic growth	X						X		X	X	X	X	X	54
Out-gassing				X	X		X					X		50, 64
Overstress		X						X	X	X	X	X	X	16, 28, 30, 34, 36, 38, 42, 44
Pattern cracking	X	X					X		X					
Pattern loading		X						X						
Peeling	X					X	X					X		
Pillowing	X		X		X				X	X				
Pinpoint corrosion	X			X		X	X			X		X		
Pitting	X	X											X	
Pitting corrosion	X	X		X	X		X			X				
Plastic hinge		X						X	X	X	X			
Point contact		X							X					

Element	Aesthetic	Strength/stress/break/rupture	Movement/distortion/strain/stability	Change in properties	Change in volume	Loss of function	Deterioration	Structural systems, foundations, and soil	Concrete, masonry, and stone	Metals	Wood and wood products	Plastic, coatings, sealants, and rubber	Glass and glazing	Case Histories (page no.)
	Aesthetic	Strength/stress/break/rupture	Movement/distortion/strain/stability	Change in properties	Change in volume	Loss of function	Deterioration	Structural systems, foundations, and soil	Concrete, masonry, and stone	Metals	Wood and wood products	Plastic, coatings, sealants, and rubber	Glass and glazing	
Ponding	X	X	X				X	X						
Pull-out		X	X						X	X	X			
Pull-through		X	X						X					
Punching shear		X						X	X					
Reaction	X	X		X			X		X	X		X	X	18
Relaxation		X						X	X					
Reversion				X		X	X					X		48
Ringworm corrosion		X		X	X		X			X				
Roll-out			X									X	X	
Rundown	X											X		46
Rupture		X						X	X	X			X	
Salt fretting	X	X					X		X					
Sand pocket		X							X					
Scaling	X				X		X		X					
Scour		X						X	X					
Segregation		X							X					
Separation		X							X					
Seizing		X				X		X		X				
Selective attack		X		X	X		X			X				
Settlement			X		X			X						
Shattering				X	X	X	X					X		
Shrinkage					X				X		X	X		22, 24, 28, 50, 52, 54, 56, 64
Side sway			X					X						
Sinkhole			X					X						
Slab curling			X		X				X					
Sliding		X							X					
Slip-stick movement			X							X				60
Slope stability failure			X					X						
Soft rot	X	X		X		X	X				X			
Spalling		X			X		X		X					36, 56, 58, 68
Spandrel volatilization	X												X	
Split	X	X					X		X		X			
Spontaneous breakage		X											X	38
Stain	X								X	X		X	X	
Staining	X								X	X		X	X	

Element	Manifestations of Distress							Materials and Systems						Reference
	Aesthetic	Strength/stress/break/rupture	Movement/distortion/strain/stability	Change in properties	Change in volume	Loss of function	Deterioration	Structural systems, foundations, and soil	Concrete, masonry, and stone	Metals	Wood and wood products	Plastic, coatings, sealants, and rubber	Glass and glazing	Case Histories (page no.)
Stratification		X							X					
Stray current corrosion		X		X	X	X	X			X				
Stress concentration		X						X	X	X	X	X	X	66
Stress corrosion cracking		X		X	X	X	X			X				
Subsidence			X					X						
Sugaring		X		X			X		X					20
Sulfate attack					X		X		X					
Swelling					X			X						68
Telegraphing	X								X					
Thermal cracking		X							X					
Thermal expansion	X	X	X		X	X		X	X	X	X	X	X	12, 14, 44, 54, 70
Thermal hysteresis		X	X	X	X		X		X					20
Thermal shock		X							X				X	20, 38
Thermo-galvanic corrosion		X		X	X	X	X			X				
Torsional buckling			X					X						
Undermining			X					X						
Uniform attack		X		X	X	X	X			X				
Vein deterioration	X			X	X		X		X					
Vortex shedding		X						X						
Walking			X						X	X		X		
Warp	X	X			X				X	X	X	X		
Water absorption	X						X		X			X		46
Water pocket		X							X					
Wax bleed	X										X	X		
Weathering	X			X			X		X	X	X	X		50
Web buckling			X					X	X					
Web crippling			X					X	X					
Web shear buckling			X					X	X					
Weld decay		X		X	X		X			X				
Wet rot	X	X		X		X	X					X		
Yield		X	X						X	X	X	X		

GLOSSARY OF FORENSIC ENGINEERING PRACTICE

Accident. An undesirable and unintentional event resulting in harm, damage, or loss.

Act of God. See the preferred term Natural Disaster.

Arbitration. The submission of a dispute to one or more impartial persons for final and binding resolution.

Baseball Arbitration. Also known as "last best offer arbitration," a special type of arbitration in which the arbitrator's decision must be selected only from the parties' final positions.

Catastrophic. Description of an unintended event resulting in loss of life, severe personal injury, or substantial property damage.

Cause. That which acts, happens, or exists in such a way that some specific thing occurs as a result; the producer of an effect or event. Causation encompasses not only failure mechanisms but also procedural errors and contributing circumstances. The following terms are often associated with cause:

- But-For Cause—an essential factor in causation, "but for" which the event would not have occurred.
- Contributing Cause—because the cause of a failure is a broad definition and already encompasses many factors, a better term would be Contributing Factor.
- Fundamental Cause—see the preferred term Proximate Cause.
- Immediate Cause—see the preferred term Proximate Cause.
- Initiating Cause—see the preferred term Proximate Cause.
- Primary Cause—see the preferred term Proximate Cause.
- Probable Cause—because of its connotation in criminology, a better term would be Most Probable Cause: the most likely set of circumstances, procedures, and failure mechanisms believed to have caused a failure.
- Procedural Cause—a nonphysical factor contributing to a failure, such as error.
- Proximate Cause—a procedural or a technical cause of failure that is deemed to be the most significant factor in producing a specific effect.
- Technical Cause—the failure mechanism that is deemed to have caused a failure.
- Triggering Cause—see the preferred term Proximate Cause.

Circumstantial Evidence. Most commonly used in criminology but also useful in forensic engineering to describe plausible but indirect evidence that cannot be positively proved to be causal.

Collapse. A structural failure resulting in the total destruction of all or a portion of a structural system, with consequential damage to the nonstructural systems.

Damage. Distress to property (including building components) caused by a Failure.

Danger. The qualitative combination of Hazard and Risk that yields the exposure of person or property to harm, injury, damage, or loss.

Defect. The nonconformity of a component with a standard or specified characteristic. Defect is used sometimes as a synonym for "failure," but the preferred meaning is to indicate only a deviation from some (perceived) standard that may, but will not necessarily, result in a failure. See also Latent Defect.

Deficient. Lacking some desirable element or characteristic.

Deviation. Divergence of the value of a quantity from a standard or reference value. Deviation is used generally to indicate a divergence from what was originally intended.

Distress. The individual or collective physical manifestations of a failure as perceivable problems, such as cracks, spalls, staining, or leakage.

Durability. The quality of maintaining satisfactory aesthetic, economic, and functional performance for the useful life of a material or system.

Evaluation. A systematic approach of obtaining information and rendering judgment about a subject. See also Examination, Investigation, and Inspection.

Event. A manifestation occurring instantaneously or over a short period of time; an incident, as opposed to a process.

Examination. A study of physical evidence. See also Evaluation, Investigation, and Inspection.

Examplar. A specimen of physical evidence of known origin, such as a reference standard specimen.

Exfiltration. Leakage out of a material or structure.

Existing Conditions. Documentable characteristics of a system that exist at a specific time, such as before an event or during an investigation.

Factor of Safety. For a particular material, the industry-recognized standard ratio of the minimum stress at which an expected failure mode (such as tensile) would develop to the maximum stress that should be used in design. As a subtle distinction, factor of safety (or sometimes "actual factor of safety") is also used to describe the ratio of the actual minimum capacity of a structure or component (taking into account all member properties, not just the material) to the maximum intended (design) service stress level.

Failure. An unacceptable difference between expected and observed performance; also, the termination of the ability of an item or system to perform an intended or required function. In forensic engineering, failure should be used only to describe physical dysfunction and not broader concepts such as business failure or errors. Not all failures are catastrophic; most involve components that do not perform as expected. The following terms are often associated with failure:

- Functional Failure—the inability of a component or system to perform an intended, essential function. It is not necessarily implied that the component or system has suffered a failure mechanism, only that it does not function as intended.

- Procedural Failure—see the preferred term Procedural Cause.
- Safety Failure—a functional failure that creates a hazard.
- Serviceability Failure—a physical failure that impairs the serviceability of a system but not its structural integrity.
- Structural Failure—a physical failure that impairs the structural integrity of a system up to and including collapse.

Failure Mechanism. An identifiable phenomenon that describes the process or defect by which an item or system suffers a particular type of failure.

Failure Mode. A description of the general type of failure experienced by a system. A broader term than failure mechanism, encompassing fundamental behavior such as shear, tension, and so forth.

Fault. Culpability. Also sometimes used to mean Defect, which would be the preferred term. Also an out-of-plane offset across a crack.

Flaw. A relatively small imperfection in a material or component. Note, however, that even small flaws can cause catastrophes if they occur in critical areas.

Forensic Engineering. The application of engineering principles to the investigation of failures or other performance problems. Forensic Engineering is often used in the resolution of disputes over the cause of and liability for failures, including providing expert testimony before a court of law or other forum.

Hazard. An attribute of a component or system that presents a threat of harm, injury, damage, or loss to person or property. See also Danger and Risk. (A hazard is not necessarily dangerous.)

Hypothesis. A proposition or set of propositions set forth as an explanation for the occurrence of a phenomenon or event, subject to confirmation by testing and used to guide an investigation.

Infiltration. Air or water leakage through a material or system and into a space that is not directly or intentionally exposed to the air or water source.

Inspection. Similar to Examination but implying more comprehensiveness or thoroughness.

Investigation. A detailed study that involves gathering and examining evidence. See also Evaluation, Examination, and Inspection.

Latent Defect. A defect that causes or has the potential to cause a failure after a relatively long period of time during which it was dormant or nonmanifest.

Linear Indication. A visible linear discontinuity in a material, especially welds, that may indicate a Crack.

Loss. The consequences of a defect or failure, expressed in terms of costs, injuries, loss of life, and so forth.

Mediation. A process in which a neutral person assists the parties to reach their own settlement of a dispute. The mediator does not have authority to make a binding decision.

Natural Disaster. One of the following generally recognized phenomena of nature resulting in destruction of property and/or personal injury: fire, flood, hurricane, major storm, tornado, hail, earthquake, avalanche, or blizzard.

Progressive Collapse. The collapse of multiple bays or floors of a structure resulting from an isolated structural failure due to a chain reaction or "domino effect."

Pseudorandom. A selection of samples from a population without bias as to the parameters being studied, intended to represent the entire population yet based on a mathematical model, such as a random number generator.

Random. A selection by chance, without mathematical sequence. When referring to sample selection methodology, see the preferred term Pseudorandom.

Rehabilitate. Extensive maintenance intended to bring a property or building up to current acceptable condition, often involving improvements.

Renovate. Make new; remodel.

Repair. To restore an item to an acceptable condition by the renewal, replacement, or mending of distressed parts.

Restore. To bring an item back to its original appearance or state.

Risk. The nonquantitative exposure of person or property to a hazard. Also, the quantitative probability that an event will occur.

Sound. Good structural condition.

Stable. Resistant to buckling or stability failure. Stable should not be used as a synonym for "good condition."

Theory. A formulation of apparent relationships or underlying principles of observed phenomena. Generally, a theory is considered valid if it explains all currently recorded data and can successfully predict the outcome of an experiment.

Annotated Bibliography of Forensic Engineering

Introduction

A printed bibliography can never be complete; new works are continually published, and additional research can uncover more references. But new technologies may finally make it possible to keep an active database of references. The Internet was designed to solve problems exactly like this; bibliographies can be continually added to and refined when stored in an electronic format. Toward that end, this annotated bibliography is a starting point. I hope that the next edition of this bibliography will be available on the World Wide Web.

My goal was to provide at least one reference for each failure mechanism discussed in this book. I have annotated certain key bibliographic entries, to assist the reader in selecting appropriate references. The entries are subdivided into the following categories:

I. Forensic Engineering Practice
II. Litigation and Expert Testimony
III. Alternative Dispute Resolution (ADR)
IV. Failure Mechanisms and Case Histories
  A. General
  B. Structural Systems, Foundations, Soil/Structure Interaction
  C. Concrete, Masonry, and Stone
  D. Metals
  E. Wood and Wood Products
  F. Plastic, Coatings, Sealants, and Rubber
  G. Glass and Glazing

The first category includes general works, which tend to be interdisciplinary. Forensic engineering often involves testimony on the findings of failure investigations before a court of law, so the second category directly applies to practice in this field. Because there is a general trend toward the use of ADR instead of litigation to resolve construction disputes, the third category is becoming increasingly important for the practice of forensic engineering. The fourth category is the broadest. The practice of forensic engineering relies heavily on understanding the causes of particular failure incidents. There are countless publications that address the performance of building materials, but those listed here are believed to be the most useful in understanding how materials can fail.

I. Forensic Engineering Practice

Avoidance of Building Failures. 1975. *Society of Architectural and Associated Technicians News*, July, pp. 16–18.

Failure Patterns and Implications. 1975. *Building Research Establishment Digest*, No. 176, Building Research Station, Garston, Watford, U.K., Apr.

Expert Systems (diagnose building failures). 1982. *RIBA Journal*, Vol. 89, p. 5, Nov.

Guide to Investigation of Structural Failures. 1979. ASCE, Research Council on Performance of Structures, New York, N.Y.

Guidelines for Failure Investigation. 1989. ASCE, Technical Council on Forensic Engineering, New York, N.Y. This publication summarizes the TCFE's recommended practice for forensic engineering.

Journal of Performance of Constructed Facilities. Technical Council on Forensic Engineering, ASCE, New York. This journal is published quarterly. It contains numerous interesting case histories of forensic investigations.

Journal of the National Academy of Forensic Engineers. NAFE, Hawthorne, N.Y. This journal, published twice per year, contains numerous interesting case histories of failure investigations.

Quality in the Constructed Project. 1990. ASCE, New York, N.Y. Although intended as a guideline for new construction, this manual is also useful in dispute resolution. Written by groups representing most of the parties to a construction project, it provides consensus interpretation of the duties of the various parties.

Shepard's Expert and Scientific Evidence Quarterly. (ongoing). Colorado Springs, Colo.

Standard Practice for Reporting Opinions of Technical Experts. ANSI/ASTM E 620-77.

Structural Failures. 1981. *Civil Engineering*, ASCE, Vol. 51, pp. 42–45, Dec.

Structural Failures: Modes, Causes, Responsibilities. 1973. ASCE, Research Council on Performance of Structures, New York, N.Y.

Technology, Law & Insurance. *Journal of the International Society for Technology, Law, and Insurance (ISTLI)*, Chapman & Hall, London.

Addleson, L. 1989. *Building Failures: A Guide to Diagnosis, Remedy and Prevention*, 2nd ed. The Architectural Press, Ltd., London, England.

Ballast, D. K. 1987. **Building Failures: Recent Book and Periodical Literature.** *Architecture Series Bibliography A 1874*, Vance Bibliographies, Monticello, Ill.

Bayazit, N. 1985. **A Study on the Diagnosis of Building Failures.** *Methods & Theories*, Vol. 19, No. 2, pp. 268–280.

Beaumont-Markland, A. 1981. **Practice: Avoiding Failure.** *RIBA Journal*, Vol. 88, p. 32, Feb.

Bell, G. R. 1985. *Failure Information Needs in Civil Engineering: Reducing Failures of Engineered Facilities.* American Society of Civil Engineers, New York.

Blockley, D., ed. 1992. *Engineering Safety.* McGraw Hill, Berkshire, U.K. Includes chapters on probabilistic risk assessment, design codes, law, bridge failures, and the psychology of risk.

Borges, J. 1991. *The Concept of Risk in Building Pathology.* CIB W86 Paper No. 7/5, Oct.

Carper, Kenneth L., ed. 1986. *Forensic Engineering: Learning from Failures.* ASCE, New York. This book contains the proceedings of a 1986 symposium.

Carper, Kenneth L., ed. 1989. *Forensic Engineering.* Elsevier Science Publishing, New York. This book is recommended for its in-depth coverage of the practice of forensic engineering.

Carper, K. L., and Feld, J. 1997. *Construction Failure.* 2nd ed. John Wiley, New York. A rich source of case histories on virtually every failure mechanism, with thorough references.

Cheetham, D. W. 1973. **Defects in Modern Buildings.** *Building*, Vol. 225, pp. 91–94, Nov. 2.

Chesson, E. 1979. **How Not to Do It!** *Proceedings of the National Conference on Engineering Case Studies*, ASCE, Washington, D.C., Mar. 28–30.

Dixon, E. Joyce. 1989. **Forensic Engineering— Ethical Practices.** *Proceedings, 1989 Annual Conference.* American Society for Engineering Education, Washington, D.C., pp. 1033–1035.

Fitzsimons, N., and D. Vannoy. 1984. **Establishing Patterns of Building Failures.** *Civil Engineering*, ASCE, Vol. 54, No. 1.

Freeman, I. L. 1974. **Failure Patterns and Implications.** *Building Defects and Failures—a Joint Building Research Establishment/Institute of Building Seminar*, Nov.

Godfrey, E. 1924. *Engineering Failures and Their Lessons.* Superior Printing Co., Akron, Ohio.

Godfrey, K. A. 1984. **Building Failures—Construction Related Problems and Solutions.** *Civil Engineering*, ASCE, Vol. 54, No. 5.

Godfrey, K. A. 1984. **Building Failures—Design Problems and Solutions.** *Civil Engineering*, ASCE, Vol. 54, No. 6, pp. 62–66.

Haines, Daniel W. 1983. **Forensic Engineering: What Role for ASCE?** *Civil Engineering*, ASCE, No. 7, pp. 53–55. The ASCE Technical Council on Forensic Engineering was formed after taking a survey of members, reported in this article, which showed overwhelming support for ASCE to take a more active role in preventing engineering failures.

Hammond, R. 1957. *Engineering Structural Failures: the Causes and Results of Failure.* New York Philosophical Library, New York, N.Y.

Janney, J. R. 1979. *Guide to Investigation of Structural Failures.* American Society of Civil Engineers, New York.

Kaminetzky, Dov. 1991. *Design and Construction Failures: Lessons from Forensic Investigations.* McGraw-Hill, New York. This book was reviewed in the February 1992 issue of the *Journal of Performance of Constructed Facilities*. A compilation of case studies, the book concludes with a discussion of forensic practice.

Khachaturian, N., ed. 1985. *Reducing Failures of Engineered Facilities.* ASCE, New York.

Leonards, G. A. 1982. **Investigation of Failures.** *Journal of the Geotechnical Engineering Division*, ASCE, Vol. 108, No. 2.

Littlemore, D. 1977. **Building Defects—Who Dunnit?** *Defects in Buildings—29th Conference of the Building Science Forum of Australia*, New South Wales Division, Sydney, p. 66.

Mathey, R. G., and J. R. Clifton. 1988. **Review of Nondestructive Evaluation Methods Applicable to Construction Materials and Structures.** *NBS Technical Note 1247*, National Bureau of Standards, June.

McKaig, T. H. 1962. *Building Failures: Case Studies in Construction and Design.* McGraw-Hill, New York.

Petroski, H. 1985. *To Engineer is Human: The Role of Failure in Successful Design.* St. Martin's Press, New York.

Porteous, W. A. 1985. **Perceived Characteristics of Building Failure: A Survey of Recent Literature.** *Architectural Science Review*, Vol. 28, June, pp. 30–40.

Rabeneck, A. 1981. **Facing up to Failure: Diagnosing and Preventing Failure.** *Architect's Journal*, Vol. 173, April 15, pp. 715–718.

Ransom, W. H. 1981. *Building Failures: Diagnosis and Avoidance*. E. & F.N. Spon, New York.

Ross, S. S. 1984. *Construction Disasters: Design Failures, Causes and Prevention*. McGraw-Hill Book Co., New York.

Sowers, George F. 1987. **Investigating Failure.** *Civil Engineering*, ASCE, No. 5, pp. 83–85. Subtitled "Pinpointing the technical cause of a failure is only one part of a successful forensic investigation."

Specter, Marvin M. 1987. **The National Academy of Forensic Engineers.** *Journal of Performance of Constructed Facilities*, ASCE, No. 8, pp. 145–149. This is a good introduction to the NAFE organization.

Tolstoy, N. 1989. **The Design of Field Investigations for Estimating the Extent of Building Failures.** *Proceedings 11th CIB Int. Congress*, Theme II, Vol. 1, June.

Waller, R. A., and Covello, V. T., eds. 1984. **Low-Probability High-Consequence Risk Analysis.** In *Advances in Risk Analysis*, Plenum Press, New York.

Wilson, F. 1984. *Building Materials Evaluation Handbook*. Van Nostrand Reinhold, New York.

II. Litigation and Expert Testimony

ACEC Liability and Litigation Report. 1988. American Consulting Engineers Council, Vol. 2, No. 6, Nov./Dec.

For the Defense. The Defense Research Institute, Inc., 750 North Lake Shore Drive, Suite 500, Chicago, Ill., 60611. A monthly publication by attorneys practicing in civil litigation defense, which is full of interesting techniques for providing effective testimony.

EXPERT: A Guide to Forensic Engineering and Service as an Expert Witness. 1987. Association of Engineering Firms Practicing in the Geosciences, Silver Spring, Md. Highly recommended for its summary of the litigation process and recommendations for the practice of forensic engineering.

Guidelines for the P.E. as a Forensic Engineer. 1985. NSPE, Alexandria, Va. This booklet is recommended for its bullet lists of guidelines for the practice of forensic engineering. It is Publication Number 1944, available from NSPE, P.O. Box 96163, Washington, D.C. 20090-6163.

Recommended Practices for Design Professionals Engaged as Experts in the Resolution of Construction Industry Disputes. Association of Engineering Firms Practicing in the Geosciences, Silver Spring, Md. These brief ethical guidelines have been adopted by at least 25 professional societies, including ASCE. It is available from ASFE, 8811 Colesville Road, Suite G106, Silver Spring, Md. 20910.

The Testifying Expert. (n.d.). LRP Publications, Horsham, Pa.

The Whole Truth and Nothing But? 1987. *Engineering News Record*. McGraw-Hill, New York, June 4, pp. 24–30.

Bachner, John P. 1988. **Facing Down the Hired Gun.** *Journal of Performance of Constructed Facilities*, ASCE, pp. 190–198.

Budinger, F. C. 1987. **Engineers In Court.** *Civil Engineering*, ASCE, No. 8, pp. 52–54. Gives tips on testifying.

Carper, Kenneth L. 1990. **Ethical Considerations for the Forensic Engineer Serving as an Expert Witness.** *Business and Professional Ethics Journal*, Vol. 9, Nos. 1 and 2, Spring–Summer, pp. 21–34. This paper is an excellent summary of ethical guidelines for the role of engineers in dispute resolutions.

Cushman, Robert F., et al. 1989. *Construction Failures*. Wiley Law Publications, New York. A textbook, with supplements published annually.

Dolan, Thomas J. 1973. **So, You Are Going To Testify As An Expert.** *ASTM Standardization News*, ASTM, West Conshohocken, Pa., Mar., pp. 30–33. Interesting historically: Professor Dolan clearly defined the ethical standards for forensic engineering more than 20 years ago.

Friedlander, Mark C. 1989. **The Design Professions: Let's Regulate Expert Witnesses.** *Civil Engineering*, ASCE, No. 4, p. 6. Friedlander makes the case for professional societies to police themselves.

Huber, Peter. 1991. **Junk Science in the Courtroom.** *Forbes*, July 8. Extreme examples of egregious expert testimony, with commentary on the syndrome of junk science.

Peters, L. W. 1989. *Construction Engineering Evidence*. Wiley Law Publications, New York. A textbook, with supplements published annually.

Postol, L. P. 1987. *A Legal Primer for Expert Witnesses.* **For The Defense.** The Defense Research Institute, Chicago, Ill., Feb. This article is designed to provide "assistance to expert witnesses so that they will feel comfortable with their duties in law suits and can maximize their assistance to attorneys."

Schwartz, M., and N. F. Schwartz. 1981. *Engineering Evidence.* McGraw-Hill Book Co., Inc., New York.

Specter, M. M. 1988. **What Does It Take To Be A Good Expert Witness?** 1988. *ASTM Standardization News,* ASTM, West Conshohocken, Pa., Feb., pp. 38–40. Subtitled, "Put the gun away, you won't need it," this is a rebuttal to a feature article on forensic engineering in another (unidentified) magazine, which implied that experts are hired guns.

Veitch, T. H. (n.d.). *The Consultant's Guide to Litigation Services: How To Be an Expert Witness.* John Wiley & Sons, New York.

III. Alternative Dispute Resolution (ADR)

Avoiding and Resolving Disputes in Underground Construction: Successful Practices and Guidelines. 1989. ASCE Technical Committee on Contracting Practices. This publication is recommended for its coverage of Dispute Review Boards (DRBs).

Electrical Design Library: Alternative Dispute Resolution (ADR) for the Construction Industry. 1987. National Electrical Contractors Association, Bethesda, Md. Gives an excellent summary of 18 ADR methods.

McManamy, Rob. 1991. **Quiet Revolution Brews for Settling Disputes.** *Engineering News Record,* McGraw-Hill, New York, Aug. 26, pp. 21–23. This cover story from *Engineering News Record,* a good status report on the industry with interesting statistics, is subtitled: "Project participants are switching from litigation to more speedy and cost-effective techniques."

Meisel, Donald D., and Walter M. Stein. 1988. **Arbitration in the Construction Industry.** *The Construction Specifier,* CSI, Alexandria, Va., Mar., pp. 120–128. An overview, including advantages and disadvantages.

Rubin, Robert A., et al. **"Baseball Arbitration"— ADR Trend of the 90s?** *The Punchlist,* American Arbitration Association, New York, pp.

3–4. A good introduction to this increasingly popular variant of arbitration.

Shanley, E. M. 1989. **A Better Way.** *Civil Engineering,* ASCE, No. 12, pp. 58–60. Subtitled: "Dispute review boards can avoid the delays, bitterness, and excessive costs of litigation or arbitration," this article makes the case for DRBs.

IV. Failure Mechanisms and Case Histories

A. General

Cracks, Movements and Joints in Buildings. 1976. *NRCC 15477, Record of the DBR Building Science Seminar 1972,* National Research Council of Canada, Division of Building Research, Ottawa.

Addleson, L., and Rice, C. 1991. *Performance of Materials in Buildings.* Butterworth-Heinemann, London, U.K.

American Society of Civil Engineers (ASCE). 1990. *Guideline for Structural Condition Assessment of Existing Buildings,* ASCE 11–90.

Eagleman, J. R., V. U. Muirhead, and N. Williams. 1975. *Thunderstorms, Tornadoes, and Building Damage.* Lexington Books, Lexington, Ma.

Lancaster, J. 1996. *Engineering Catastrophes.* Abington Publishing, Cambridge, England.

Litvan, G. G. 1980. *Freeze-Thaw Durability of Porous Building Materials.* NRCC 18638, National Research Council of Canada, Division of Building Research, Ottawa.

Nicastro, D. H. **Failures.** *Construction Specifier,* CSI, Alexandria, Va. Monthly column, January 1994 to present. Each column presents a case history of a particular failure from a variety of construction disciplines.

Portland Cement Association (PCA). 1982. **Building Movement and Joints.** EB086.01B, Portland Cement Association, Skokie, Ill.

Schild, E., et al. 1978. *Structural Failure in Residential Buildings, Volume 1: Flat Roofs, Roof Terraces, Balconies.* John Wiley & Sons, New York.

Schild, E., et al. 1978. *Structural Failure in Residential Buildings, Volume 3.* John Wiley & Sons, New York.

Schild, E., et al. 1981. *Structural Failure in Residential Buildings, Volume 2: External Walls and Openings.* John Wiley & Sons, New York.

Schild, E., et al. 1981. *Structural Failure in Residential Buildings, Volume 4: Internal Walls,*

Ceilings, and Floors. Granada Publishing Co., London.

Schlager, N. 1995. *Breakdown: Deadly Technological Disasters.* Visible Ink Press, Detroit. Includes many concise case histories of building construction failures, in addition to other industries.

Simpson, J. W., and P. J. Horrobin, eds. 1970. *The Weathering and Performance of Building Materials.* Wiley-Interscience, New York.

Stevens, A. J. 1991. *BRE Building Defect Database.* Building Research Establishment, Garston, Watford, U.K., BRE note N25/91, Feb.

Stockbridge, J. G. 1981. **Cladding Failures— Lack of a Professional Interface.** *Journal of the Technical Councils,* Vol. 105, No. 2.

Trechsel, H. R., ed. 1994. *Moisture Control in Buildings.* ASTM Manual Series MNL 18, ASTM, West Conshohocken, Pa.

United States Department of Commerce, National Bureau of Standards. 1977. **Investigation of Skyline Plaza Collapse in Fairfax County, Virginia.** *NIST (formerly NBS) Building Science Series NBSIR 94,* U.S. Government Printing Office, Washington, D.C., Feb.

United States Department of Commerce, National Bureau of Standards. 1979. **Investigation of Construction Failure of Reinforced Concrete Cooling Tower at Willow Island, West Virginia.** *NIST (formerly NBS) Building Science Series NBSIR 78-1578,* U.S. Government Printing Office, Washington, D.C., Nov.

United States Department of Commerce, National Bureau of Standards. 1981. **Investigation of Construction Failure of Harbour Cay Condominium in Cocoa Beach, Florida.** *NIST (formerly NBS) Building Science Series NBSIR 81-2374,* U.S. Government Printing Office, Washington, D.C., Sept.

United States Department of Commerce, National Bureau of Standards. 1982. **Investigation of the Kansas City Hyatt Regency Walkways Collapse.** *NIST (formerly NBS) Building Science Series NBSIR 82-2465,* U.S. Government Printing Office, Washington, D.C., May.

B. Structural Systems, Foundations, Soil/Structure Interaction

Lessons from Dam Accidents, USA. 1975. Committee on Failures and Accidents to Large Dams, of the United States Committee on Large Dams, ASCE, New York.

Allen, H. G., and P. S. Bulson. 1980. *Background to Buckling.* McGraw-Hill, New York.

Das, B. M. 1994. *Principals of Geotechnical Engineering.* 3rd ed. PWS Publishing Co., Boston.

Fintel, M., S. K. Ghosh, and H. Iyengar. 1986. *Column Shortening in Tall Structures—Prediction and Compensation.* EB108.01D, Portland Cement Association, Skokie, Ill.

Holts, R. D., and W. D. Kovacks. 1981. *An Introduction to Geotechnical Engineering.* Prentice-Hall, Englewood Cliffs.

Irvine, H. M. 1981. *Cable Structures.* MIT Press.

LePatner, B. B., and S. M. Johnson. 1982. *Structural and Foundation Failures: A Casebook for Architects, Engineers and Lawyers.* McGraw-Hill, New York.

McCormac, J. C. 1989. *Structural Steel Design: LRFD Method.* HarperCollins Publishers, New York.

Murillo, J. A. 1987. **The Scourge of Scour.** *Civil Engineering,* ASCE, No. 7, pp. 66–69.

Russell, H. G., and W. G. Corley. 1977. *Time-Dependent Behavior of Columns in Water Tower Place.* RD052.01B, Portland Cement Association, Skokie, Ill.

Shepherd, R., and J. D. Frost. 1995. *Failures in Civil Engineering: Structural, Foundation, and Geoenvironmental Case Studies.* Committee on Education, Technical Council on Forensic Engineering, ASCE, New York.

Smith, E. A., and H. Epstein. 1980. **Hartford Coliseum Roof Collapse: Structural Collapse Sequence and Lessons Learned.** *Civil Engineering,* ASCE, Vol. 50, No. 4.

Sowers, G. B., and G. F. Sowers. 1970. *Introductory Soil Mechanics and Foundations.* 3rd ed. MacMillan Company, New York.

Sputo, T. 1993. *Sinkhole Damage to Masonry Structure.* Journal of Performance of Constructed Facilities, Vol. 7, No. 1, Feb.

Terzaghi, K. 1950. *Mechanisms of Landslides.* Geological Society of America, Engineering Geology (Berkley) Volume, pp. 83–123.

Thornton, C. H. 1982. **Lessons Learned from Recent Long-Span Roof Failures.** *ACI Seminar on Lessons from Failures of Concrete Buildings,* Boston, Ma., Apr. 6.

Trahair, N. S. 1993. *Flexural-Torsional Buckling of Structures.* CRC Press, Boca Raton.

Varnes, D. J. 1978. *Slope Movement Types and Processes; Landslides: Analysis and Control.* National Academy of Sciences, Washington, D.C., Special Report 176, pp. 11–33.

Whittaker, B. N., and D. J. Reddish. 1989. *Subsi-*

dence: Occurrence, Prediction, and Control. Elsevier Science Publishers, B.V., New York.

Wierzbicki, T., and N. Jones. 1989. *Structural Failure.* John Wiley & Sons, Inc., New York.

C. Concrete, Masonry, and Stone

GFRC Recommended Practice. 3rd ed. 1993. Precast/Prestressed Concrete Institute.

Seminar Course Manual: Lessons from Failures of Concrete Buildings. (n.d.). American Concrete Institute, Detroit, Mich.

Masonry: Materials, Properties, and Performances. 1982. ASTM STP 778, American Society for Testing and Materials, West Conshohocken, Pa.

Abo-Qaidis, S. A., and I. L. Al Qadi. 1995. **Rigid Pavement Joint Sealant Effectiveness in Reducing Chloride Intrusion.** *Science and Technology of Building Seals, Sealants, Glazing and Waterproofing: Fourth Volume,* ASTM STP 1243, D. Nicastro, ed., ASTM, West Conshohocken, Pa.

American Concrete Institute (ACI). 1972. **Control of Cracking in Concrete Structures.** *Journal ACI,* Vol. 69, No. 12, pp. 717–753.

American Concrete Institute (ACI). 1979. *Guide for Making a Condition Survey of Concrete in Service.* ACI Committee 201, ACI 201.1R-68, reaffirmed 1979; American Concrete Institute, Detroit, Mich.

American Concrete Institute (ACI). 1990. *Cement and Concrete Terminology.* ACI 116R-90, American Concrete Institute, Detroit, Mich.

Allen, D. E. 1978. *Damage to Brick and Stone Veneer on Tall Buildings.* NRCC Building Practice Note 7, National Research Council of Canada, Division of Building Research, Ottawa.

Amrhein, E. 1978. **Progressive Collapse of Masonry Structures.** *Proceedings of the North American Masonry Conference.*

Beall, C. A. 1993. *Masonry Design and Deailing: for Architects, Engineers, and Contractors.* McGraw-Hill, New York.

Brick Institute of America (BIA). 1994. *Technical Notes on Brick Construction.* (periodically updated). Brick Institute of America, Reston, Va.

Bortz, S. A., B. Erlin, and C. B. Monk, Jr. 1988. *Some Field Problems with Thin Veneer Building Stones.* New Stone Technology, Design, and Construction for Exterior Wall Systems, ASTM STP 996, B. Donaldson, ed., ASTM, West Conshohocken, Pa., pp. 11–31.

Chana, P. S. 1991. **Punching Shear in Concrete Slabs.** *Structural Engineer,* Vol. 69, No. 15.

Chin, I. R., J. P. Stecich, and B. Erlin. 1990. **A Thin-Stone Veneer Primer.** *Architectural Record,* Vol. 178, No. 7, June.

Cohen, J. M., and P. J. M. Monteiro. 1991. **Durability and Integrity of Marble Cladding: A State-of-the-Art Review.** *Journal of Performance of Constructed Facilities,* ASCE, Vol. 5, No. 2, May, pp. 113–124.

Day, R. W. 1995. **Damage to Concrete Flatwork from Sulfate Attack.** *Journal of Performance of Constructed Facilities,* ASCE, Vol. 9, No. 4, Nov.

Drysdale, R. G., A. A. Hamid, and L. R. Baker. 1994. *Masonry Structures: Behavior and Design.* Prentice-Hall, Englewood Cliffs.

Feld, J. 1964. *Lessons from Failures of Concrete Structures.* American Concrete Institute, Detroit, Mich.

Grimm, C. T. 1982. *Masonry Failure Investigations.* Masonry: Materials, Properties, and Performances, ASTM STP 778, American Society for Testing and Materials, West Conshohocken, Pa.

Grimm, C. T. 1975. **Design for Differential Movement in Brick Walls.** *Journal of the Structural Division,* ASCE, Nov.

Henshell, J. 1990. **Moisture Related Problems in Masonry Parapets.** *Proceedings of the Fifth North American Conference,* Vol. IV, University of Illinois, Urbana, Ill.

Hooker, K. A. 1995. **Stone Anchoring.** *The Construction Specifier,* CSI, Alexandria, Va., Sept., pp. 49–54.

Hua, W. G., M. C. Griffith, and R. F. Warner. 1989. *Diagnosis and Assessment of Defective Concrete Structures.* Report No. R83, University of Adelaide, Dept. of Civil Engineering, Adelaide, Australia.

Hua, W. G., M. C. Griffith, and R. F. Warner. 1990. *Diagnostic Procedures for Defective Concrete Buildings.* Second National Structural Engineering Conference, Adelaide, South Australia, Oct. 3–5. Prepared for the ASCE Research Council on Performance of Structures. Lists failures by project type, structure type, and material. Covers general principles of an investigation, including field operations, testing, and analysis.

Lawrence, C. L. 1995. **Delayed Ettringite Formation: An Issue?** In *Materials Science of Concrete IV,* Jan Skalny, ed., The American Ceramic Society, Inc., Westerville, Ohio.

Lin, T. Y., and N. H. Burns. 1976. *Design of*

Prestressed Concrete Structures. 3rd ed. John Wiley & Sons, New York.

Marek, C. R. 1991. *The Aggregate Handbook.* National Stone Association, Washington, D.C.

Marusin, S. L. 1994. **Concrete Failure Caused by Delayed Ettringite Formation Case Study.** *3rd CANMET/ACI National Conference on Durability of Concrete.*

McCormac, J. C. 1993. *Design of Reinforced Concrete.* HarperCollins College Publishers.

McDougle, E. A. 1995. **GFRC Comes of Age.** *The Construction Specifier,* CSI, Alexandria, Va. Dec.

Michaud, P. 1996. **Design Guidelines for Concrete Anchoring.** *The Construction Specifier,* CSI, Alexandria, Va., Dec.

Mindess, S., and J. F. Young. 1981. *Concrete.* Prentice-Hall, Englewood Cliffs.

Monk, C. J. 1980. **Masonry Facade and Paving Failures.** *Proceedings of the Second Canadian Masonry Symposium,* Ottawa, Canada, June.

Murdock, L. J., K. M. Brook, and J. D. Dewar. 1991. *Concrete: Materials and Practice.* 6th ed. Edward Arnold, London.

Nawy, E. G. 1996. *Reinforced Concrete: A Fundamental Approach.* 3rd ed. Prentice-Hall, Englewood Cliffs.

National Ready Mixed Concrete Association (NRMCA). 1978. *What, Why & How? Dusting Concrete Surfaces.* National Ready Mixed Concrete Association.

Oesterle, Schultz, and Gilkin. 1990. *Design Considerations for GFRC Facades.* ACI SP-124, pp. 157–182.

Page, A. W., W. Samarasinghe, and A. W. Hendry. 1980. **The Failure of Masonry Shear Walls.** *The International Journal of Masonry Construction,* Vol. 1, No. 2.

Page, A. W., W. Samarasinghe, and A. W. Hendry. 1982. **The In-Plane Failure of Masonry—A Review.** *Proceedings of the BCS,* Load-Bearing Brickwork (7), Sept.

Pryor, C. A. 1992. **Alkali Silica Reactions.** *Stone Review,* June, p. 10.

Rollings, R. S. 1993. **Curling Failures of Steel-Fiber-Reinforced Concrete Slabs.** *Journal of Performance of Constructed Facilities,* ASCE, Vol. 7, No. 1, Feb.

Ropke, J. C. 1982. *Concrete Problems: Causes and Cures.* McGraw-Hill Book Co., Inc., New York.

Springenschmid, R. 1994. **Preface.** *Proceedings of the RILEM Symposium.*

Swamy, R. N. 1992. **Alkali-Aggregate Reactions in Concrete: Material and Structural Implications.** In **CANMET: Advances in Concrete Technology,** V. M. Malhotra, ed., Minister of Supply and Services, Canada.

Widhalm, C., E. Tschegg, and W. Eppensteiner. 1996. **Anisotropic Thermal Expansion Causes Deformation of Marble Claddings.** *Journal of Performance of Constructed Facilities,* ASCE, Vol. 10, No. 1, pp. 5–10.

Widhalm, C., E. Tschegg, and W. Eppensteiner. 1997. **Acoustic Emission and Anisotropic Thermal Expansion When Heating Marble.** *Journal of Performance of Constructed Facilities,* ASCE, Vol. 11, No. 1, pp. 35–40.

D. Metals

American Institute of Steel Construction (AISC). 1986. *Manual of Steel Construction: Load Factor and Resistance Design.* 1st ed. American Institute of Steel Construction, Chicago, Ill.

Bordass, W., J. Charles, and D. Farrell. 1990. **Corrosion and Decay in Sheet Metal Roofing.** *Proceedings of the Second International Conference on Building Pathology,* Sept.

Bosich, J. F. 1970. *Corrosion Prevention for Practicing Engineers.* Barnes & Noble.

Bradford, S. A. 1993. *Corrosion Control.* Van Nostrand Reinhold, New York.

Craig, B. D. 1989. *Handbook of Corrosion Data.* ASM International, Metals Park, Ohio.

Douville, J. A. 1989. *Corrosion Technology: An Information Sourcebook.* Hemisphere Publishing Co., New York.

Evans, U. R. 1963. *An Introduction to Metallic Corrosion.* 2nd ed. Edward Arnold Ltd., London, U.K.

Fisher, J. W. 1984. *Fatigue and Fracture in Steel Bridges: Case Studies.* Wiley-Interscience, New York.

Heidersbach, R. H. Jr. 1985. *Corrosion of Metals in Concrete and Masonry Buildings.* Paper 258, Corrosion 85, Mar. 25–26, NACE, Tex.

Mattsson, E. 1989. *Basic Corrosion Technology for Scientists and Engineers.* John Wiley & Sons, New York.

Pludek, V. R. 1977. **Design and Corrosion Control.** John Wiley & Sons, New York.

Raymond, L., ed. 1988. *Hydrogen Embrittlement: Prevention and Control.* ASTM, West Conshohocken, Pa.

Rosenberg, A., C. M. Hansson, and C. Andrade. 1989. **Mechanisms of Corrosion of Steel in Concrete.** In *Materials Science Concrete I,* Jan

Skalny, ed., The American Ceramic Society, Westerville, Ohio.

Salmon, C. G., and J. E. Johnson. 1990. *Steel Structures.* 3rd ed. HarperCollins Publishers.

Scott, P., and M. Davies. 1992. **Microbiologically Induced Corrosion.** *Civil Engineering,* ASCE, May.

Speller, F. N. 1951. *Corrosion, Causes, and Prevention.* 3rd ed. McGraw-Hill, New York.

Tam, C. K., and S. F. Stiemer. 1996. **Development of Bridge Corrosion Cost Model for Coating Maintenance.** *Journal of Performance of Constructed Facilities,* ASCE, May.

Uhlig, H. H., ed. 1955. *The Corrosion Handbook.* John Wiley & Sons, New York.

Von Fraunhofer, J. A. 1974. *Concise Corrosion Science.* Portcullis Press, Ltd., London.

E. Wood and Wood Products

Breyer, D. E. 1993. *Design of Wood Structures.* 3rd ed. McGraw-Hill, New York.

Singh, J., and N. White. 1997. **Timber Decay in Buildings: Pathology and Control.** *Journal of Performance of Constructed Facilities,* Vol. 11, No. 1, Feb., pp. 3–12.

F. Plastic, Coatings, Sealants, and Rubber

Science and Technology of Building Seals, Sealants, Glazing, and Waterproofing. ASTM STP 1168 (1st Vol., 1991), 1200 (2nd Vol., 1992), 1254 (3rd Vol., 1993), 1243 (4th Vol., 1994), 1271 (5th Vol., 1995), and 1286 (6th Vol., 1996), ASTM, West Conshohocken, Pa.

Cash, C. G. 1992. **Shattering PVC Roof Membranes.** *Progressive Architecture,* Feb.

Croce, S. *Detection Procedure for Diagnosing the Causes of Flat Roof Failures.* CIB W86 Paper No. 3/17.

Gish, L., ed. 1989. *Building Deck Waterproofing.* ASTM STP 1084, ASTM, West Conshohocken, Pa.

Grassie, N., and G. Scott. 1984. *Polymer Degradation and Stabilisation.* Cambridge UP.

Joy, F. A. 1963. **Premature Failure of Built-Up Roofing.** *Better Building Report 5,* Pennsylvania State University, College of Engineering.

McCampbell, B. H. 1992. *Problems in Roofing Design.* Butterworth, Heinemann, Boston.

Nicastro, D. H. 1997. **Premature Failure of Sealants.** *The Construction Specifier,* CSI, Apr.

National Roofing Contractors Association (NRCA). 1990. *Shattering of Aged Unreinforced PVC Membranes.* Joint document of the National Roofing Contractors Association and the Single Ply Roofing Institute. NRCE, Chicago, Ill.

G. Glass and Glazing

Brungs, M. P., and X. Y. Sugeng. 1995. **Some Solutions to the Nickel Sulphide Problem in Toughened Glass.** *Glass Technology,* Vol. 36, No. 4, Aug., pp. 107–110.

FGMA. 1990. *Glazing Manual.* Glass Association of North America (GANA), Topeka, Kans.

Vild, D. J. 1987. **Why Glass Goes Bad.** *Exteriors,* Spring, p. 10.